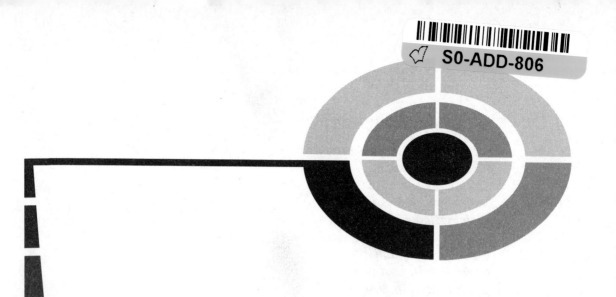

Electronics Demystified

Demystified Series

Advanced Statistics Demystified
Algebra Demystified
Anatomy Demystified
Astronomy Demystified
Biology Demystified
Business Statistics Demystified
Calculus Demystified
Chemistry Demystified
College Algebra Demystified
Differential Equations Demystified
Earth Science Demystified
Electronics Demystified
Everyday Math Demystified
Geometry Demystified
Math Word Problems Demystified
Physics Demystified
Physiology Demystified
Pre-Algebra Demystified
Pre-Calculus Demystified
Project Management Demystified
Robotics Demystified
Statistics Demystified
Trigonometry Demystified

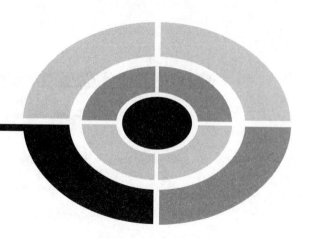

Electronics Demystified

STAN GIBILISCO

McGRAW-HILL
New York   Chicago   San Francisco   Lisbon   London
Madrid   Mexico City   Milan   New Delhi   San Juan
Seoul   Singapore   Sydney   Toronto

The McGraw·Hill Companies

Library of Congress Cataloging-in-Publication Data

Gibilisco, Stan.
    Electronics demystified/Stan Gibilisco.
        p.   cm.
    Includes bibliographical references and index.
    ISBN 0-07-143493-3
    1. Electronics—Popular works. I. Title.

  TK7819 G35   2004
  621.381—dc22   2004053169

2 3 4 5 6 7 8 9 0 DOC/DOC 0 1 0 9 8 7 6 5 4

ISBN 0-07-143493-3

The sponsoring editor for this book was Judy Bass and the production supervisor was Pamela A. Pelton. It was set in Times Roman by Keyword Publishing Services Ltd. The art director for the cover was Margaret Webster-Shapiro. Cover design by Handel Low.

Printed and bound by RR Donnelley.

 This book was printed on recycled, acid-free paper containing a minimum of 50% recycled, de-inked fiber.

McGraw-Hill books are available at special quantity discounts to use as premiums and sales promotions, or for use in corporate training programs. For more information, please write to the Director of Special Sales, McGraw-Hill Professional, Two Penn Plaza, New York, NY 10121-2298. Or contact your local bookstore.

To Tony, Samuel, and Tim from Uncle Stan

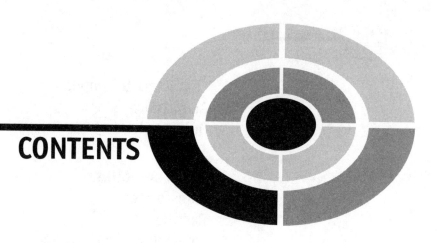

CONTENTS

Preface   xiii

PART ONE: **FUNDAMENTAL CONCEPTS**   1

CHAPTER 1   **Direct Current**   3
The Nature of DC   3
Ohm's Law   7
Resistive Networks   11
DC Magnetism   23
Quiz   28

CHAPTER 2   **Alternating Current**   31
Frequency and Waveform   31
More Definitions   39
Phase Relationships   43
Utility Power Transmission   49
Quiz   53

CHAPTER 3   **Impedance**   56
Introducing j 56
Inductive Reactance   60
Current and Voltage in RL Circuits   63
Capacitive Reactance   68
Current and Voltage in RC Circuits   71

	Characteristic Impedance	79
	Admittance	82
	Quiz	88
CHAPTER 4	**Power Supplies**	**91**
	Transformers	91
	Rectifiers	94
	Filtering and Regulation	99
	Protection of Equipment	103
	Electrochemical Power Sources	106
	Specialized Power Supplies	113
	Quiz	116
Test: Part One		**118**
PART TWO:	**WIRED ELECTRONICS**	**129**
CHAPTER 5	**Semiconductor Diodes**	**131**
	The P-N Junction	131
	Signal Applications	136
	Oscillation and Amplification	146
	Diodes and Radiant Energy	147
	Quiz	151
CHAPTER 6	**Transistors and Integrated Circuits**	**153**
	The Bipolar Transistor	153
	Basic Bipolar Transistor Circuits	160
	The Field-Effect Transistor	163
	The MOSFET	169
	Basic FET Circuits	172
	Integrated Circuits	176
	Quiz	181

Contents

CHAPTER 7	**Signal Amplifiers**	**184**
	Amplification Factor	184
	Basic Amplifier Circuits	187
	Amplifier Classes	189
	Efficiency and Drive	192
	Audio Amplification	196
	RF Amplification	199
	Quiz	204
CHAPTER 8	**Signal Oscillators**	**206**
	RF Oscillators	206
	Oscillator Stability	211
	Crystal-Controlled Oscillators	214
	AF Oscillators	219
	Quiz	222
Test: Part Two		**225**
PART THREE:	**WIRELESS ELECTRONICS**	**235**
CHAPTER 9	**Radio-Frequency Transmitters**	**237**
	Modulation	237
	Analog-to-Digital Conversion	250
	Image Transmission	251
	Quiz	255
CHAPTER 10	**Radio-Frequency Receivers**	**258**
	Simple Designs	258
	The Modern Receiver	261
	Pre-detector Stages	264
	Detectors	266
	Audio Stages	272

Television Reception   273
Specialized Modes   276
Quiz   280

CHAPTER 11   **Telecommunications**   **282**
Networks   282
Satellites   287
Personal Communications Systems   289
Hobby Communications   293
Lightning   296
Security and Privacy   299
Quiz   306

CHAPTER 12   **Antennas**   **309**
Radiation Resistance   309
Half-Wave Antennas   312
Quarter-Wave Antennas   314
Loop Antennas   317
Ground Systems   320
Gain and Directivity   322
Phased Arrays   325
Parasitic Arrays   329
UHF and Microwave Antennas   332
Feed Lines   336
Safety Issues   337
Quiz   338

Test: Part Three   **340**

Final Exam   **351**

APPENDIX 1   **Answers to Quiz, Test,
and Exam Questions**   **368**

Contents

APPENDIX 2   **Symbols Used in
Schematic Diagrams**   371

Suggested Additional References   388

Index   389

PREFACE

This book is for people who want to get acquainted with the concepts of elementary electronics without taking a formal course. It can serve as a supplemental text in a classroom, tutored, or home-schooling environment. It should also be useful for career changers who want to become familiar with basic electronics.

This course is for beginners, and is limited to elementary concepts. The treatment is mostly qualitative. There's some math here, but it does not go deep. If you want a more comprehensive treatment of electricity and electronics following completion of this book, *Teach Yourself Electricity and Electronics*, also published by McGraw-Hill, is recommended. That book goes deeper into the math, and also explores a few exotic applications.

This book contains many practice quiz, test, and exam questions. They are all multiple-choice, and are similar to the sorts of questions used in standardized tests. There is an "open-book" quiz at the end of every chapter. You may (and should) refer to the chapter texts when taking them. When you think you're ready, take the quiz, write down your answers, and then give your list of answers to a friend. Have the friend tell you your score, but not which questions you got wrong. Stick with a chapter until you get most of the answers correct.

This book is divided into three sections called "parts." At the end of each section is a multiple-choice test. Take these tests when you're done with the respective sections and have taken all the chapter quizzes. The section tests are "closed-book," but the questions are not as difficult as those in the quizzes. A satisfactory score is 75%. There is a final exam at the end of this course. It contains questions from all the chapters. Take this exam when you have finished all sections and tests. A satisfactory score is at least 75%.

With the section tests and the final exam, as with the quizzes, have a friend tell you your score without letting you know which questions you missed. That way, you will not subconsciously memorize the answers. You might want to take each test, and the final exam, two or three times. When you have gotten a score that makes you happy, you can check to see where your knowledge is strong and where it is not so keen.

Answers to the quizzes, section tests, and the final exam are in an appendix at the end of the book. A table of schematic symbols is included in a second appendix.

I recommend that you complete one chapter a week. An hour or two daily ought to be enough time for this. When you're done with the course, you can use this book as a permanent reference.

Suggestions for future editions are welcome.

Stan Gibilisco

Acknowledgments

Illustrations in this book were generated with *CorelDRAW*. Some of the clip art is courtesy of Corel Corporation.

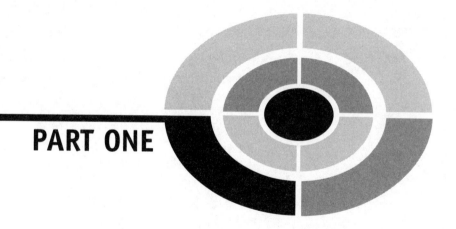

PART ONE

Fundamental Concepts

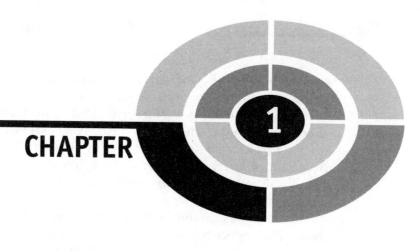

Direct Current

Direct current (DC) is a flow of electricity, or *current*, that takes place in an unchanging direction. This is what distinguishes direct current from *alternating current* (AC), in which the current goes back and forth.

The Nature of DC

Figure 1-1 (on page 7) illustrates four simple graphs of current versus time. The graphs at A, B and C depict DC because the current always flows in the same direction, even though the amplitude (intensity) might change with time. The rendition at D is not DC, because the direction of current flow does not stay the same all the time.

WHAT IS CURRENT?

Current is a measure of the rate at which electrical *charge carriers*, usually *electrons*, flow. A current of one *ampere* (1 A) represents one *coulomb* (6.24×10^{18}) of charge carriers per second past a given point. Often, current is specified in terms of *milliamperes*, abbreviated mA, where $1\,\text{mA} = 0.001\,\text{A}$. You will also sometimes read or hear of *microamperes* (μA), where $1\,\mu\text{A} = 10^{-6}\,\text{A} = 0.001\,\text{mA}$. Once in a while, you will read or hear about *nanoamperes* (nA), where $1\,\text{nA} = 0.001\,\mu\text{A} = 10^{-9}\,\text{A}$. A current of a few milliamperes through your body will give you a shock; 50 mA will jolt you severely; 100 mA can kill you if it flows through your heart.

There are some circuits in which extremely large currents flow. Imagine a thick, solid copper bar placed directly across the output terminals of a large electric generator. The generator can drive huge numbers of electrons through the bar every second. In other circuits, the flow of current is relatively small (although the number of electrons per second may still be quite large). In microcomputers, a few nanoamperes will suffice for the execution of complicated electronic processes.

RESISTANCE

Resistance is the opposition that a component, device, or circuit offers to the flow of electric current. The standard unit of resistance is the *ohm*, symbolized Ω. Other common resistance units include the *kilohm* (abbreviated kΩ or k), where $1\,\text{k} = 1{,}000\,\Omega$, and the *megohm* (abbreviated MΩ or M), where $1\,\text{M} = 1{,}000\,\text{k} = 10^{6}\,\Omega$.

The nano-, micro-, milli-, kilo-, and mega- prefixes are just a few of the *prefix multipliers* used by scientists to denote large and small quantities. Table 1-1 is a complete list. It's good to know most of these by heart. They all represent powers of 10. (There is another list that represents powers of 2, and these are used in the computer industry. They're a little different than the ones shown here.)

Electrical components, devices, or circuits always have at least a little bit of resistance. Materials with low resistance are good electrical *conductors*. In some electronic applications, materials are selected on the basis of how large their resistance is, and the larger the better. These materials make good electrical *insulators*.

In a practical electronic circuit, the resistance of a particular component might vary depending on the conditions under which it is operated. A transistor, for example, might have high resistance some of the time, and low resistance at other times. The high/low resistance fluctuation in a transistor can take place thousands, millions, or billions of times each second.

Table 1-1 Prefix multipliers and their abbreviations.

Prefix	Symbol	Multiplier
yocto-	y	$\times$ 0.000000000000000000000001 (10^{-24})
zepto-	z	$\times$ 0.000000000000000000001 (10^{-21})
atto-	a	$\times$ 0.000000000000000001 (10^{-18})
femto-	f	$\times$ 0.000000000000001 (10^{-15})
pico-	p	$\times$ 0.000000000001 (10^{-12})
nano-	n	$\times$ 0.000000001 (10^{-9})
micro-	μ	$\times$ 0.000001 (10^{-6})
milli-	m	$\times$ 0.001
centi-	c	$\times$ 0.01
deci-	d	$\times$ 0.1
(none)	—	$\times$ 1
deka-	da or D	$\times$ 10
hecto-	h	$\times$ 100
kilo-	K or k	$\times$ 1000
mega-	M	$\times$ 1,000,000 (10^{6})
giga-	G	$\times$ 1,000,000,000 (10^{9})
tera-	T	$\times$ 1,000,000,000,000 (10^{12})
peta-	P	$\times$ 1,000,000,000,000,000 (10^{15})
exa-	E	$\times$ 1,000,000,000,000,000,000 (10^{18})
zetta-	Z	$\times$ 1,000,000,000,000,000,000,000 (10^{21})
yotta-	Y	$\times$ 1,000,000,000,000,000,000,000,000 (10^{24})

ELECTROMOTIVE FORCE

The standard unit of *electromotive force* (EMF) is the *volt*. An EMF of one volt, (1 V), when it is placed across a component with a resistance of 1 Ω, drives a current of 1 A through that component. Does the term EMF seem sophisticated or esoteric? If so, think of it as "electrical pressure." Most people call it *voltage*. Common voltage units, besides the volt itself, include the *microvolt* (μV), where $1\,\mu V = 0.000001\,V = 10^{-6}\,V$, the *millivolt* (mV), where $1\,mV = 0.001\,V$, the *kilovolt* (abbreviated kV), where $1\,kV = 1000\,V$, and the *megavolt* (abbreviated MV), where $1\,M = 1,000\,kV = 10^{6}\,V$.

It is possible to have an EMF without a flow of current. This is *static electricity*. The term *static* means *unmoving*. Static electricity exists just before a lightning stroke occurs. It also exists across the terminals of a battery when there is nothing connected. Charge carriers move only if a conductive path is provided. The fact that an EMF is large does not guarantee that it will drive a large current through a circuit. A good example is your body after walking around on a carpet. This produces an EMF of many volts. Although the idea of thousands of volts building up on your body can be scary, there are not many electrons involved. The "static" shock you sometimes receive when you touch something metallic is small because the current is small.

If there are many coulombs of available electric charge carriers, a moderate voltage can drive a lethal current through the human body. This is why it is dangerous to repair some electrical and electronic devices when they are powered-up. A utility power supply can pump an unlimited number of coulombs through your body. If the conductive path is through the heart, electrocution can occur.

SOURCES OF DC

Typical sources of DC include the power supplies for computers and radios, *electrochemical cells* and batteries, and *photovoltaic cells* and panels. The intensity, or amplitude, of DC might fluctuate with time, and this fluctuation might be periodic. In some such cases the DC has an AC component superimposed on it (as in Fig. 1-1B). A source of DC is sometimes called a *DC generator*.

Batteries and various other sources of DC produce a constant voltage. This is called *pure DC*, and can be represented by a straight, horizontal line on a graph of voltage versus time (as in Fig. 1-1A). In some instances the value of DC voltage pulsates or oscillates rapidly with time, in a manner similar to the changes in an AC wave. This is called *pulsating DC*.

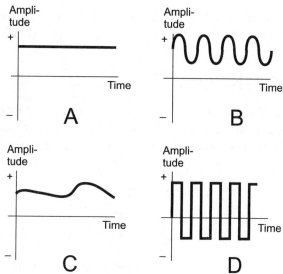

Fig. 1-1.   Examples of DC waveforms (A, B, C), and a non-DC waveform (D).

SCHEMATIC DIAGRAMS

This book contains diagrams of simple electrical and electronic circuits. When individual components (such as resistors, batteries, capacitors, coils, and transistors) must be shown in a diagram, special symbols are used to represent them. The wires connecting the components are shown as solid lines.

If you have not worked very much with *schematic diagrams*, they can be bewildering. In order to get familiar with the symbols, spend some time studying Appendix 2 at the end of this book. It is a comprehensive table of electronic symbols used in schematic diagrams.

In schematic diagrams, connections between components, which physically are wires or foil strips on printed-circuit boards, are illustrated as solid straight lines. These lines are almost always oriented either "north-south" (vertically) or "east-west" (horizontally). When two lines in a diagram cross each other, the diagram does not indicate that the conductors are connected, unless there is a dot at the point where they cross.

Ohm's Law

Any DC circuit can be simplified to three parts: a source of voltage, a set of current-carrying conductors, and an overall resistance. The voltage provided

to the circuit is symbolized E (or sometimes V); the current through the circuit is symbolized I; the resistance of the circuit is symbolized R.

STATEMENT OF THE LAW

The interdependence among voltage, current, and resistance is one of the most fundamental rules of electrical circuits. It is called *Ohm's Law*, named after the scientist who supposedly first expressed it. Three formulas denote this law:

$$E = IR$$
$$I = E/R$$
$$R = E/I$$

You need only remember the first equation in order to derive the others. The easiest way to remember it is to learn the abbreviations E for EMF, I for intensity (of current), and R for resistance, and then remember that they appear in alphabetical order with the equals sign after the E. You can also imagine them inside a triangle as shown in Fig. 1-2.

If the initial quantities are given in units other than volts, amperes, and ohms, you must convert to these standard units, and then calculate. After that, you can convert the units back again to whatever you like.

CURRENT, VOLTAGE, AND RESISTANCE CALCULATIONS

Refer to Fig. 1-3. The circuit shown in this diagram consists of a variable DC generator, a *voltmeter* for measuring EMF, some wire, an *ammeter* for measuring current, and a calibrated, wide-range variable resistor called a *potentiometer*. The following three problems (and solutions) demonstrate how Ohm's Law can be used to calculate unknown current, voltage, and resistance in a DC circuit.

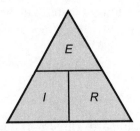

Fig. 1-2.   The "*EIR* triangle" can serve as a memory aid.

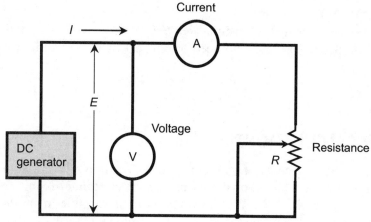

Fig. 1-3.   Circuit for demonstrating Ohm's Law.

PROBLEM 1-1
Suppose the DC generator produces 10 V, and that the potentiometer is set to a value of 10 Ω. What is the current?

SOLUTION 1-1
The current can be found by the formula $I = E/R$. Plug in the values for E and R; they are both 10, because the units are given in volts and ohms. Therefore:

$$I = 10/10 = 1\,\text{A}$$

PROBLEM 1-2
Suppose the potentiometer in Fig. 1-3 is set to 100 Ω, and the measured current is 10 mA. What is the voltage across the resistance?

SOLUTION 1-2
Use the formula $E = IR$ to find the voltage. First, convert the current to amperes: 10 mA = 0.01 A. There is no need to convert the resistance, because it is given in ohms already. Then multiply:

$$E = 0.01 \times 100 = 1\,\text{V}$$

PROBLEM 1-3
Suppose both the voltmeter and ammeter scales in Fig. 1-3 are visible, but the potentiometer is uncalibrated. If the voltmeter reads 24 V and the ammeter shows 3 A, what is the resistance of the potentiometer?

SOLUTION 1-3

This can be found using the formula $R = E/I$. We are given the initial units in volts and amperes, so there's no need to convert them. Dividing, we obtain:

$$R = 24/3 = 8\,\Omega$$

POWER CALCULATIONS

You can calculate the electrical *power* dissipated by the resistance in a DC circuit such as that shown in Fig. 1-3. The standard unit of power is the *watt*, symbolized W. When expressed as a variable in mathematical equations, power is symbolized P. Use the following formula to calculate power based on voltage and current:

$$P = EI$$

The power in watts is the product of the voltage in volts and the current in amperes.

If you know I and R, but do not know E, you can get the power P by means of this formula:

$$P = I^2 R$$

You can also get the power if you are not given the current directly. Suppose you know only the voltage and the resistance. Then:

$$P = E^2/R$$

The above three formulas hold only when units are expressed in watts, volts, amperes, and ohms. Before performing power calculations, always be sure the quantities have been converted to these standard units. Then you can be sure your result will be expressed in watts.

PROBLEM 1-4

Suppose that the voltmeter in Fig. 1-3 reads 12 V and the ammeter shows 50 mA. Find the power dissipated in the resistor.

SOLUTION 1-4

First, convert the current to amperes, getting $I = 0.050$ A. We are given the voltage directly as $E = 12\,V$. Thus:

$$P = EI = 12 \times 0.050 = 0.60\,\text{W}$$

PROBLEM 1-5

If the resistance is 999 Ω and the voltage source delivers 3 V, find the power dissipated in the resistor.

SOLUTION 1-5

We are given the resistance directly in ohms and the voltage directly in volts, so there is no need to convert either of these units. The calculation proceeds simply:

$$P = 3^2/999 = 9/999 = 0.009 \text{ W}$$

This can be expressed as 9 milliwatts (mW), where 1 mW = 0.001 W.

PROBLEM 1-6

If the resistance is 0.47 Ω and the current is 680 mA, find the power dissipated in the resistor.

SOLUTION 1-6

First, convert the current to amperes; $I = 0.680$ A. We are already given the resistance in ohms, so there is no need to convert it. Therefore:

$$P = 0.680^2 \times 0.47 = 0.22 \text{ W}$$

Resistive Networks

Combinations of resistive components, also called *resistances*, are used in DC circuits to regulate the current and control the voltage.

RESISTANCES IN SERIES

When you place resistances in series, their values in ohms add up directly to get the total resistance, as long as all the resistances are in the same units. If some of the resistances are in kilohms and others are in megohms, you must convert all the resistances to the same units (ohms, kilohms, or megohms, as you prefer) before adding them.

RESISTANCES IN PARALLEL

When resistances are placed in parallel, they behave differently than they do in series. In general, if you have a resistor of a certain value and you place other resistors in parallel with it, the overall resistance decreases.

One way to look at resistances in parallel is to consider them as conductive components, or *conductances*, instead. In parallel, conductances add directly, just as resistances add in series. If you change all the ohmic values to *siemens* (the standard unit of conductance), you can add these figures up and convert the final answer back to ohms.

The symbol for conductance is G. The abbreviation for siemens is S. The conductance G of a component, in siemens, is related to the resistance R, in ohms, by the formulas

$$G = 1/R$$
$$R = 1/G$$

To find the net resistance of a set of three or more resistances in parallel, follow this procedure:

- Convert all resistance values to ohms
- Find the reciprocal of each resistance; these are the respective conductances in siemens
- Add up the individual conductance values; this is the net conductance of the combination
- Take the reciprocal of the net conductance; this is the net resistance, in ohms, of the parallel combination

When there are several resistances in parallel and their values are all equal, the total resistance is equal to the resistance of any one component, divided by the number of components.

PROBLEM 1-7
Suppose the following resistances are hooked up in series with each other: $R_1 = 112\,\Omega$, $R_2 = 470\,\Omega$, and $R_3 = 680\,\Omega$. What is the net series resistance, R, of this combination?

SOLUTION 1-7
The resistance, as determined across the whole series combination, is found by adding the values:

$$R = R_1 + R_2 + R_3$$
$$= 112 + 470 + 680 = 1262\,\Omega$$

PROBLEM 1-8
Consider five resistors in parallel. Call them R_1 through R_5, and call the total resistance R as shown in Fig. 1-4. Let $R_1 = 100\,\Omega$, $R_2 = 200\,\Omega$, $R_3 = 300\,\Omega$,

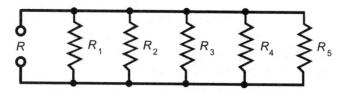

Fig. 1-4. Five resistances in parallel, R_1 through R_5, give a total resistance R as discussed in the text.

$R_4 = 400\,\Omega$, and $R_5 = 500\,\Omega$. What is the net parallel resistance, R? Round off the answer to the nearest tenth of an ohm.

SOLUTION 1-8

Converting the resistances to conductance values, we get $G_1 = 1/100 = 0.01\,\text{S}$, $G_2 = 1/200 = 0.005\,\text{S}$, $G_3 = 1/300 = 0.00333\,\text{S}$, $G_4 = 1/400 = 0.0025\,\text{S}$, and $G_5 = 1/500 = 0.002\,\text{S}$. Adding these gives:

$$G = 0.01 + 0.005 + 0.00333 + 0.0025 + 0.002$$
$$= 0.02283\,\text{S}$$

The total resistance, rounded off to the nearest tenth of an ohm, is therefore:

$$R = 1/G = 1/0.02283 = 43.8\,\Omega$$

DIVISION OF POWER

When combinations of resistances are hooked up to a source of voltage, they draw current. You can determine how much current they will draw by calculating the total resistance of the combination, and then considering the whole network as a single resistor.

If the resistances in the network all have the same ohmic value, the power from the source is evenly distributed among the resistances, whether they are hooked up in series or in parallel. If the resistances in the network do not all have identical ohmic values, the power will be divided up unevenly among them. Then the power in any particular resistor must be calculated by determining either the current through that resistor or the voltage across it, and using the applicable DC power formula.

RESISTANCES IN SERIES-PARALLEL

A group of resistors, all having identical ohmic values, can be connected together in parallel sets of series components, or in series sets of parallel components. These are called *series-parallel resistive networks*. By connecting a lot of identical resistors together this way, the total power-handling capacity can be greatly increased over that of a single resistor.

The total resistance of a series-parallel network is the same as the value of any one of the resistors as long as the resistances are all identical, and are connected in a network called an *n-by-n matrix*. That means, when *n* is a whole number, there are *n* parallel sets of *n* resistors in series (Fig. 1-5A), or else there are *n* series sets of *n* resistors in parallel (Fig. 1-5B). These two arrangements, although geometrically different, produce the same practical result.

Engineers and technicians sometimes use *n-by-n* matrices to advantage to obtain resistive components with large power-handling capacity. In such a matrix, all the resistors must have the same power-dissipation rating. Then the combination of *n-by-n* resistors will have n^2 times that of a single resistor. A 3-by-3 series-parallel matrix of 2 W resistors can handle $3^2 \times 2 = 18$ W, for example. A 10×10 array of 1 W resistors can dissipate 100 W.

Non-symmetrical series-parallel networks, made up from identical resistors, can increase the power-handling capability over that of a single resistor. But in these cases, the total resistance is not the same as the value of the single resistors. The overall power-handling capacity is multiplied by the total number of resistors whether the network is symmetrical or not,

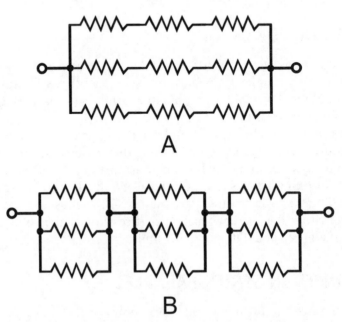

A

B

Fig. 1-5.   Series-parallel combinations. At A, sets of series resistors are connected in parallel. At B, sets of parallel resistors are connected in series.

provided all the resistors have the same resistance and the same power-handling capacity.

CURRENTS THROUGH SERIES RESISTANCES

In a series DC circuit, the current at any given point is the same as the current at any other point. This is true no matter what the components are, and regardless of whether or not they all have the same resistance.

If the various components have different resistances in a series DC circuit, some of them consume more power than others. In case one of the components shorts out, the current through the whole chain increases, because the overall resistance of the string goes down. If a component opens, the current drops to zero at every point in a series circuit, because the flow of charge carriers is interrupted.

VOLTAGES ACROSS SERIES RESISTANCES

In a series circuit, the voltage is divided up among the components. The sum total of the voltages across each resistance is equal to the supply voltage. This is always true, no matter how large or how small the resistances, and whether or not they are all of the same value.

The voltage across any resistor in a series combination is equal to the product of the current and the resistance. Remember to use volts, ohms, and amperes when making calculations. In order to find the current in the circuit, you need to know the total resistance and the supply voltage. Then $I = E/R$. First find the current in the whole circuit; then find the voltage across any particular resistor.

VOLTAGES ACROSS PARALLEL RESISTANCES

In a parallel DC circuit, the voltage across each component is equal to the supply or battery voltage. The current drawn by any particular component depends on its individual resistance. In this sense, the components in a parallel-wired circuit work independently, as opposed to the series-wired circuit in which they interact.

If any one branch of a parallel circuit is taken away, the conditions in the other branches remain the same. If new branches are added, assuming the power supply can deliver enough current, conditions in previously existing branches are not affected.

By the way, here's another definition for you to remember. When a DC voltage exists across a particular component or set of components, or between two specific points in a circuit, that voltage is sometimes called a *potential difference*.

CURRENTS THROUGH PARALLEL RESISTANCES

Refer to the schematic diagram of Fig. 1-6. The total parallel resistance in the circuit is R. The battery voltage is E. The current in branch n, containing resistance R_n, is measured by ammeter A, and is called I_n.

The sum of the currents through all the resistances in the circuit is equal to the total current, I, drawn from the source. The current is divided up in the parallel circuit, just as voltage is divided up in a series circuit.

PROBLEM 1-9

Consider a specific case of the scenario shown in Fig. 1-6. Imagine there are 10 resistors in parallel. One of these resistors, R_n, has a value of $100\,\Omega$.

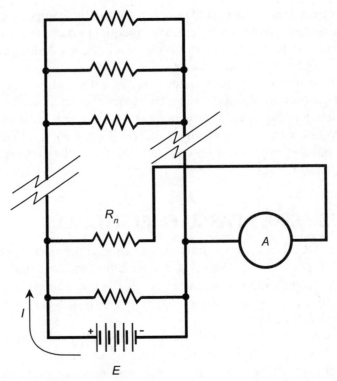

Fig. 1-6.   Analysis of current in a parallel DC circuit.

Suppose the current through R_n is 0.2 A. What is the voltage, E, of the battery?

SOLUTION 1-9
The voltage of the battery can be found by determining the voltage across R_n. In a parallel DC circuit, the entire supply or battery voltage appears across each and every one of the resistors. This is a simple case of Ohm's Law:

$$E = IR$$
$$= 0.2 \times 100$$
$$= 20\,V$$

PROBLEM 1-10
Suppose that in the situation described in the previous problem, all 10 of the resistors have values of 100 Ω. If the battery supplies 20 V, what is the total current drawn from it by the parallel set of resistors? How much power is demanded from the battery?

SOLUTION 1-10
The total current is the sum of the currents through each of the resistors. We already know that this current is 0.2 A in the case of R_n. Because all the resistors have the same value, the current through each one is the same. Therefore, the total current drawn from the battery is $0.2 \times 10 = 2\,A$. We can determine the power demanded from the battery using the formula for power in terms of voltage and current:

$$P = EI$$
$$= 20 \times 2$$
$$= 40\,W$$

POWER DISTRIBUTION IN SERIES CIRCUITS

When calculating the power dissipated by a particular resistor R_n in a circuit containing n resistors in series, first determine the current, I, that the circuit carries. Then it is easy to calculate the power P_n, based on the formula $P_n = I^2 R_n$.

The total power, or *wattage*, dissipated in a series circuit is equal to the sum of the wattages dissipated in each resistor. In this sense, the distribution of power in a series circuit is like the distribution of the voltage.

POWER DISTRIBUTION IN PARALLEL CIRCUITS

When resistances are wired in parallel, the current is not the same in each resistance. An easier method to find the power P_n, dissipated by a particular resistor R_n, is to use the formula $P_n = E^2/R_n$, where E is the voltage of the battery or power supply. This voltage is the same across every resistor in a parallel circuit.

In a parallel circuit, the total dissipated wattage is equal to the sum of the wattages dissipated by the individual resistances. All types of DC circuits, parallel or series, share this trait. All power in any DC circuit must be accounted for. Power cannot appear from out of nowhere, nor can it vanish into nowhere.

KIRCHHOFF'S LAWS

The behavior of currents and voltages in DC circuits is governed by two rules known as *Kirchhoff's laws*. They are sometimes called *Kirchhoff's First Law* and *Kirchhoff's Second Law*. So we don't get confused about which one is first and which one is second, let's name them according to their descriptions of current and voltage.

Kirchhoff's Law for Current: The current going into any point in a DC circuit is the same as the current going out of that point. This is true no matter how many branches lead into or out of the point (Fig. 1-7A). A qualitative way of saying this is that current can never appear from nowhere or disappear into nowhere.

Kirchhoff's Law for Voltage: The sum of all the voltages from a fixed point in a DC circuit all the way around and back to that point from the opposite direction, taking polarity into account, is always zero (Fig. 1-7B). Stated another way, voltage can never appear from nowhere or disappear into nowhere.

VOLTAGE DIVIDERS

Resistances in series produce potential differences that can be tailored to meet certain needs. A circuit of series-connected resistances, deliberately set up to produce specific DC voltages at various points, is called a *voltage divider*.

In the design of a voltage divider, the resistances should be as small as possible, without causing too much current drain on the power supply. This ensures that the voltages are stable under varying conditions. Any system connected to a voltage divider, and that relies on that divider for its operation,

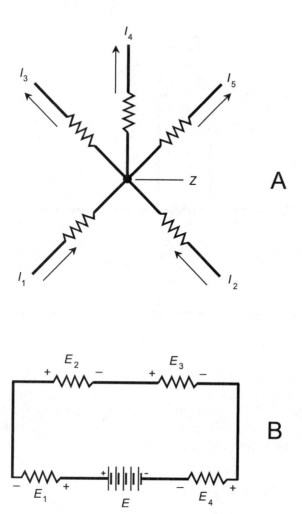

Fig. 1-7. At A, Kirchhoff's Law for Current. The current entering point Z is equal to the current leaving point Z. In this case, $I_1 + I_2 = I_3 + I_4 + I_5$. At B, Kirchhoff's Law for Voltage. In this case, $E + E_1 + E_2 + E_3 + E_4 = 0$, taking polarity into account.

affects the voltages. The lower the resistances in the divider, the less pronounced is this effect. Ideally, the total series resistance of the divider should be a small fraction of the resistance of the system operating from that divider.

Figure 1-8 illustrates the principle of voltage division. The individual resistances are R_1, R_2, R_3, ..., R_n. The total resistance is $R = R_1 + R_2 + R_3 + \cdots + R_n$, because the resistors are connected in series. The battery voltage is E, so the current drawn from the battery is equal to $I = E/R$. At the points P_1, P_2, P_3, ..., P_n, the voltages, with respect to the negative terminal of the battery, are E_1, E_2, E_3, ..., E_n. The highest voltage, E_n, is equal to the

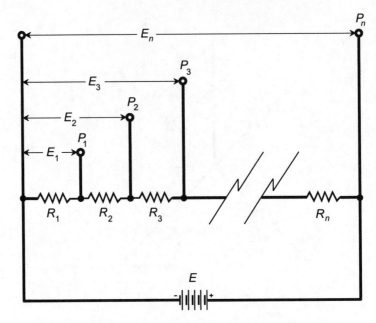

Fig. 1-8.   General arrangement for a voltage divider circuit.

supply voltage, E. All the other voltages are less than E, so $E_1 < E_2 < E_3 < \cdots < E_n = E$.

The voltages at the various points increase according to the sum total of the resistances up to each point, in proportion to the total resistance, multiplied by the supply voltage. That is:

$$E_1 = E(R_1/R)$$
$$E_2 = E(R_1 + R_2)/R$$
$$E_3 = E(R_1 + R_2 + R_3)/R$$
$$\downarrow$$
$$E_n = E(R_1 + R_2 + R_3 + \cdots + R_n)/R = E(R/R) = E$$

These voltages can also be found by multiplying the circuit current I, which is equal to E/R, by the incremental resistances, as follows:

$$E_1 = IR_1$$
$$E_2 = I(R_1 + R_2)$$
$$E_3 = I(R_1 + R_2 + R_3)$$
$$\downarrow$$
$$E_n = I(R_1 + R_2 + R_3 + \cdots + R_n) = IR = E$$

PROBLEM 1-11

Consider a voltage divider constructed according to the circuit design shown in Fig. 1-8, in which there are five resistors with the following values, in order:

$$R_1 = 300 \, \Omega$$
$$R_2 = 250 \, \Omega$$
$$R_3 = 200 \, \Omega$$
$$R_4 = 150 \, \Omega$$
$$R_5 = 100 \, \Omega$$

If the battery has voltage $E = 100 \, \text{V}$, what are E_1 through E_5 at the points P_1 through P_5?

SOLUTION 1-11

The total resistance R of the series combination is:

$$R = R_1 + R_2 + R_3 + R_4 + R_5$$
$$= 300 + 250 + 200 + 150 + 100$$
$$= 1000 \, \Omega$$

Therefore, the current I through the circuit is:

$$I = E/R$$
$$= 100/1000$$
$$= 0.1 \, \text{A}$$

This current flows through each of the resistors. We can determine the voltages E_1 through E_5 by multiplying I by the incremental resistances at the points P_1 through P_5 (the total resistances between those points and the negative battery terminal):

$$E_1 = 0.1 \times 300$$
$$= 30 \, \text{V}$$

$$E_2 = 0.1 \times (300 + 250)$$
$$= 0.1 \times 550$$
$$= 55 \, \text{V}$$

$$E_3 = 0.1 \times (300 + 250 + 200)$$
$$= 0.1 \times 750$$
$$= 75 \, \text{V}$$

$$E_4 = 0.1 \times (300 + 250 + 200 + 150)$$
$$= 0.1 \times 900$$
$$= 90\,\mathrm{V}$$

$$E_5 = 0.1 \times (300 + 250 + 200 + 150 + 100)$$
$$= 0.1 \times 1000$$
$$= 100\,\mathrm{V}$$

PROBLEM 1-12

How much power, in watts, does each resistor dissipate in the above situation?

SOLUTION 1-12

Let's call these power levels P_1 through P_5. (Note that here, P refers to the word "power," not the word "point.") The current I is the same through each of the resistors, and we know that $I = 0.1\,\mathrm{A}$. The square of the current, I^2, is 0.01 A. Thus:

$$P_1 = I^2 R_1$$
$$= 0.01 \times 300$$
$$= 3\,\mathrm{W}$$

$$P_2 = I^2 R_2$$
$$= 0.01 \times 250$$
$$= 2.5\,\mathrm{W}$$

$$P_3 = I^2 R_3$$
$$= 0.01 \times 200$$
$$= 2\,\mathrm{W}$$

$$P_4 = I^2 R_4$$
$$= 0.01 \times 150$$
$$= 1.5\,\mathrm{W}$$

$$P_5 = I^2 R_5$$
$$= 0.01 \times 100$$
$$= 1\,\mathrm{W}$$

DC Magnetism

Whenever the atoms in a material are aligned in a certain way, a *magnetic field* exists. A magnetic field can also be caused by the motion of electric charge carriers such as electrons, either in a wire or in free space.

MAGNETIC FLUX

Scientists and engineers consider magnetic fields to be comprised of *flux lines*. These lines are imaginary, not material objects; and they are usually curves, not straight lines. Each line represents a certain quantity of magnetism. The intensity of a magnetic field is determined according to the number of flux lines passing through a certain unit area, such as a square centimeter (cm^2) or a square meter (m^2).

A magnetic field is considered to originate at the north magnetic pole, and to terminate at the south magnetic pole. Every line of flux connects the two poles. The greatest *flux density*, or magnetic field strength, around a bar magnet or a horseshoe magnet is near the poles, where the lines converge. Around a current-carrying wire, the greatest flux density is near the wire.

MAGNETIC FIELD STRENGTH

The overall quantity, or amount, of magnetism produced by a magnetic object or device is measured in *webers*, abbreviated Wb. A smaller unit, the *maxwell* (Mx), is used if the magnetic field is weak. One weber is equivalent to $100,000,000\,(10^8)$ Mx. Conversely, $1\,Mx = 0.00000001\,(10^{-8})$ Wb.

Flux density is a more useful expression for magnetic effects than the overall quantity of magnetism. A flux density of one *tesla* (T) is equal to one weber per square meter ($1\,Wb/m^2$). In this case, we can imagine each flux line as $1\,Wb$. A flux density of one *gauss* (G) is equal to one maxwell per square centimeter ($1\,Mx/cm^2$). In this case, we can imagine each flux line as $1\,Mx$. One gauss is equivalent to $0.0001\,(10^{-4})$ T. Conversely, $1\,T = 10,000\,(10^4)\,G$.

When defining the properties of an *electromagnet*, which is a wire coil deliberately designed to produce a strong magnetic field, another unit of magnetism is often employed: the *ampere-turn* (At). A wire, bent into a circle and carrying $1\,A$ of current, produces $1\,At$ of *magnetomotive force*. The *gilbert* (Gb) is occasionally used to express magnetomotive force in electromagnets. One ampere-turn is equal to $0.4\pi\,Gb$, or about $1.257\,Gb$. Conversely, $1\,Gb$ is equal to $10/(4\pi)\,At$, or about $0.7958\,At$. The value of π, a mathematical constant, is approximately 3.14159.

AMPERE'S LAW

The intensity of the magnetic field at any specific point near an electrical conductor carrying DC is directly proportional to the current. For a straight wire, the magnetic flux takes the form of concentric cylinders centered on the wire. A cross-sectional view of this situation is shown in Fig. 1-9. The wire is perpendicular to the page, so it appears as a point, and the flux lines appear as circles.

Physicists define *theoretical current* as flowing from the positive pole to the negative (even though electrons flow from the negative pole to the positive). With this in mind, suppose the theoretical current depicted in Fig. 1-9 flows out of the page toward you. According to *Ampere's Law*, the direction of the magnetic flux is counterclockwise in this situation.

Ampere's Law is sometimes called the *right-hand rule*. If you hold your right hand with the thumb pointing out straight and the fingers curled, and then point your thumb in the direction of the plus-to-minus theoretical current flow in a straight wire, your fingers curl in the direction of the magnetic flux. Similarly, if you orient your right hand so your fingers curl in the direction of the magnetic flux, your thumb points in the direction of the theoretical current.

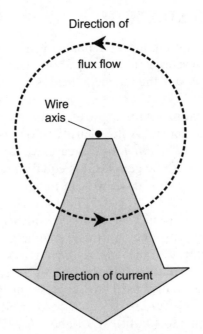

Fig. 1-9.   Ampere's Law for a current-carrying conductor.

PERMEABILITY

Permeability is a measure of the way in which a substance affects magnetic flux density. A material that concentrates magnetic lines of flux is called *ferromagnetic*. Such substances can be "magnetized." Iron is a good example, although some specialized alloys are more ferromagnetic than pure iron. A substance that dilates, or dilutes, magnetic flux is called *diamagnetic*. Wax, dry wood, bismuth, and silver are substances that have this effect. The dilution of magnetic flux in diamagnetic materials is nowhere near as pronounced as the concentration of the flux in ferromagnetic materials.

Permeability is measured on a scale relative to a vacuum (or dry air, which behaves like a vacuum with respect to magnetic fields). This "empty" medium is also known as *free space*. Free space is assigned permeability 1.

Suppose you place a rod-shaped iron core inside a long, cylindrical coil (called a *solenoidal coil*) with the intent of making an electromagnet. The flux density in the core increases by a factor of approximately 60 to 8000 (depending on the purity of the iron) compared with the flux density inside the coil when the iron is not there. This means that the permeability of iron is anywhere from about 60 (impure) to about 8000 (refined). If you use certain alloys as the core materials in electromagnets, you can increase the flux density inside the core material by as much as 1,000,000 times. The permeability factors of some common materials are shown in Table 1-2.

Diamagnetic substances are used to keep magnetic objects apart while minimizing the interaction between them. Non-ferromagnetic metals, such as copper and aluminum, conduct electric current well, but magnetic flux poorly. They can be used for *electrostatic shielding*, a means of allowing magnetic fields to pass through while blocking electric fields. The *electric flux* is in effect shorted out by a sheet or screen made of copper, but the magnetic flux passes right on through.

THE RELAY

A *relay* (Fig. 1-10) makes use of an electromagnet to allow remote-control switching. The movable lever, called the *armature*, is held to one side by a spring when there is no current flowing through the electromagnet. Under these conditions, terminal X is connected to terminal Y, but not to terminal Z. When a sufficient current is applied, the armature is pulled over to the other side. This disconnects terminal X from terminal Y, and connects X to Z.

Table 1-2   Permeability figures for some common materials.

Substance	Permeability (approx.)
Aluminum	Slightly more than 1
Bismuth	Slightly less than 1
Cobalt	60–70
Ferrite	100–3000
Free space	1
Iron	60–100
Iron, refined	3000–8000
Nickel	50–60
Permalloy	3000–30,000
Silver	Slightly less than 1
Steel	300–600
Super permalloys	100,000 to 1,000,000
Wax	Slightly less than 1
Wood, dry	Slightly less than 1

A *normally closed relay* completes the circuit when there is no current flowing in its electromagnet, and breaks the circuit when current flows. A *normally open relay* is just the opposite. It completes the circuit when current flows in the coil, and breaks the circuit when there is no current. Some relays have several sets of contacts. Some are designed to remain in one state (either with current or without) for a long time, while others are built to switch several times per second.

There is a limit to how fast a mechanical relay can switch between states, and the contacts eventually become corroded or wear out. For these reasons, relays are not used much anymore; they have been largely replaced by semiconductor switches. But in some high-current or high-voltage systems,

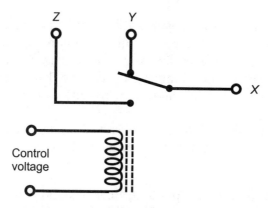

Fig. 1-10. Schematic representation of a simple relay. See the text for discussion of terminals
X, *Y*, and *Z*.

relays are still used because they are more "forgiving" of unexpected voltage
spikes than are semiconductor devices.

MAGNETIC TAPE

Magnetic tape was once used for home entertainment, especially hi-fi music and
home video. It is still found in some older audio and video equipment. The tape
consists of millions of ferromagnetic particles glued to a flexible, nonmagnetic
strip. A fluctuating magnetic field, produced by the recording head, polarizes
these particles. As the field changes in strength next to the recording head, the
tape passes by at a constant, controlled speed. This produces regions in which
the particles are polarized in either direction.

When the tape is run at the same speed in the playback mode, the magnetic
fields around the individual particles cause a fluctuating field that is detected
by the pickup head. This field has the same pattern of variations as the
original field from the recording head.

The data on a magnetic tape can be distorted or erased by external
magnetic fields. Extreme heat can also result in loss of data, and possible
physical damage to the tape.

MAGNETIC DISK

The principle of the *magnetic disk*, on the micro scale, is the same as that of
the magnetic tape. The information is stored in digital form; that is, there are
only two different ways that the particles are magnetized. This results in
almost perfect, error-free storage.

On a larger scale, the disk works differently than tape because of the difference in geometry. On a tape, the information is spread out over a long span, and some bits of data are far away from others. But on a disk, no two bits are farther apart than the diameter of the disk. This means that data can be stored and retrieved much more quickly onto, or from, a disk than is possible with a tape.

The same precautions should be observed when handling and storing magnetic disks, as are necessary with magnetic tape.

PROBLEM 1-13
Suppose a length of wire is wound 10 times around a plastic hoop, and a current of 0.5 A flows through the wire. What is the resulting magnetomotive force in ampere-turns? In gilberts?

SOLUTION 1-13
To find the magnetomotive force in ampere-turns, multiply the current in amperes by the number of turns. This gives the result $0.5 \times 10 = 5$ At. To convert this to gilberts, multiply by 1.257; this comes out to $5 \times 1.257 = 6.285$ Gb.

Quiz

Refer to the text in this chapter if necessary. Answers are in the back of the book.

1. Suppose a wire, running straight up and down, carries a steady direct current. Suppose a flat sheet of paper is placed in a horizontal plane, so it is perpendicular to the wire, with the wire passing through the paper. The magnetic flux intersects the paper
    (a) in points along a straight line.
    (b) in concentric circles.
    (c) in parallel lines.
    (d) nowhere.

2. What is the DC resistance between the two end terminals of the series-parallel network shown in Fig. 1-11?
    (a) 10 Ω
    (b) 20 Ω
    (c) 30 Ω
    (d) 60 Ω

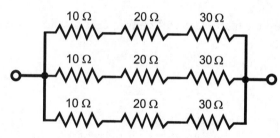

Fig. 1-11.   Illustration for Quiz Questions 2 through 5.

3. Suppose a battery that supplies 6 V is connected to the end terminals of the series-parallel resistive network shown in Fig. 1-11. What is the current through the 10 Ω resistor at the upper left?
    (a) 0.06 A
    (b) 0.1 A
    (c) 0.6 A
    (d) 1 A

4. Suppose a battery that supplies 12 V is connected to the end terminals of the series-parallel resistive network shown in Fig. 1-11. What is the voltage across the 20 Ω resistor at the center?
    (a) 2 V
    (b) 3 V
    (c) 4 V
    (d) 6 V

5. Suppose a battery that supplies 24 V is connected to the end terminals of the series-parallel resistive network shown in Fig. 1-11. What is the power dissipated by the 30 Ω resistor at the lower right?
    (a) 12 W
    (b) 7.5 W
    (c) 6 W
    (d) 4.8 W

6. According to Ohm's Law, if the resistance in a circuit remains constant and the voltage across that resistance decreases, then
    (a) the current through the resistance increases.
    (b) the current through the resistance decreases.
    (c) the current through the resistance reverses direction.
    (d) the current through the resistance drops to zero.

7. The rate at which electrical charge carriers flow is measured in
   (a) watts.
   (b) volts.
   (c) ohms.
   (d) amperes.

8. A ferromagnetic material
   (a) cannot conduct an electrical current.
   (b) must carry a large current.
   (c) concentrates magnetic lines of flux.
   (d) has negative magnetic polarity.

9. In order to be considered DC, an electrical current
   (a) must always flow in the same direction.
   (b) must have negative polarity.
   (c) cannot consist of charge carriers.
   (d) must be constant.

10. The data on a magnetic disk can be erased by
    (a) bright light.
    (b) a strong magnetic field.
    (c) a high resistance.
    (d) any of the above.

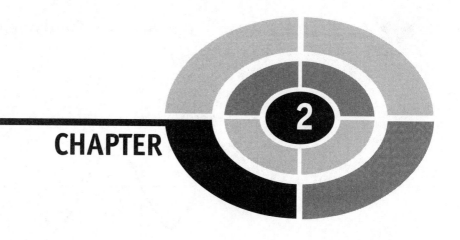

Alternating Current

In alternating current (AC), the direction of current, or the polarity of voltage, reverses at a constant rate. In most AC utility circuits in the United States, it happens 60 times per second. In radio transmitters and receivers, an AC wave can reverse polarity billions of times per second.

Frequency and Waveform

In a *periodic AC wave*, the current or voltage, as a function of time, cycles back and forth indefinitely. The length of time between a specific point in a cycle and the same point in the next cycle is called the *period* of the wave (Fig. 2-1).

A WIDE RANGE

Theoretically, the period of an AC wave can range anywhere from a minuscule fraction of a second to many years. In electronics, engineers and technicians usually work with signals in the *audio frequency* (AF) or

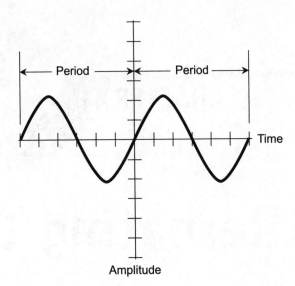

Fig. 2-1. A sine wave. The period is the length of time required for one cycle to be completed.

radio frequency (RF) ranges. Currents at AF reverse polarity from about 20 times per second to 20,000 times per second. These figures represent the lowest and highest frequencies, respectively, that human beings can hear when AC is converted into sound waves. Currents at RF reverse polarity from a few thousand times per second up to many billions of times per second.

The period of an AC wave, when defined in seconds, is denoted by the uppercase italicized letter *T*. The *frequency*, denoted by the lowercase italicized letter *f*, of a wave is the reciprocal of the period when *f* is expressed in cycles per second and *T* is expressed in seconds. Mathematically, we can state this as either of the following two equations:

$$f = 1/T$$
$$T = 1/f$$

The standard unit of frequency is the *hertz*, abbreviated Hz, which represents one complete AC cycle per second. Higher frequencies are given in *kilohertz* (kHz), *megahertz* (MHz) or *gigahertz* (GHz). The relationships are:

$$1\,\text{kHz} = 1000\,\text{Hz}$$
$$1\,\text{MHz} = 1000\,\text{kHz} = 10^6\,\text{Hz}$$
$$1\,\text{GHz} = 1000\,\text{MHz} = 10^9\,\text{Hz}$$

Sometimes an even larger unit, the *terahertz* (THz), is used. This is a million-million (10^{12}) hertz.

Some AC waves consist of energy entirely at one frequency, called the *fundamental frequency*. These are *pure sine waves*. Often, there are components, known as *harmonics*, at whole-number multiples of the fundamental frequency. There might also be components at unrelated frequencies. Some waves are extremely complex, consisting of hundreds, thousands or even infinitely many different *component frequencies*.

SINE WAVE

The classical AC wave has a sine-wave, or *sinusoidal*, shape. The waveform in Fig. 2-1 is a sine wave. Any AC wave that consists of a single frequency has a perfect sine-wave shape. The converse of this is also true: perfectly sinusoidal AC contains only one component frequency.

In practice, a wave can be so close to a sine wave that it looks exactly like the sine function on an oscilloscope, when in fact it is not. (An oscilloscope shows time on the horizontal axis and amplitude on the vertical axis, and it portrays amplitude as a function of time. For this reason, it is called a *time-domain* display.) Waveform imperfections are often too small to see on a conventional oscilloscope display. Utility AC in the United States has an almost perfect sine-wave shape, but it isn't perfect. The imperfections are responsible for *electromagnetic noise* that is radiated by some AC power lines.

SQUARE WAVE

On an oscilloscope, a perfect *square wave* looks like a pair of parallel, dashed lines, one having positive polarity and the other having negative polarity (Fig. 2-2A). When a square wave is illustrated, the transitions can be shown as vertical lines (Fig. 2-2B).

A square AC wave might have equal negative and positive peaks. Then the *absolute amplitude* of the wave is constant at a certain voltage, current, or power level. Half of the time the amplitude is a certain positive number of volts, amperes, or watts, and the other half of the time the amplitude is the same number of negative volts, amperes, or watts. This is called a *symmetrical square wave*.

In an *asymmetrical square wave*, the positive and negative amplitudes are not the same. For example, the positive peak might be +5 V and the negative peak −3 V. It's also possible for the length of time during which the amplitude is positive to differ from the length of time the amplitude is negative. In this sort of situation the wave is not truly square, but is described by the more general term *rectangular wave*.

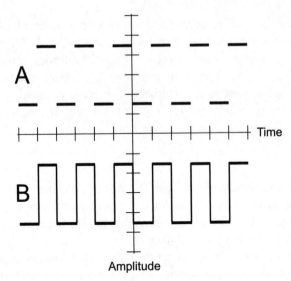

Fig. 2-2.   At A, a theoretically perfect square wave. At B, the transitions are shown as vertical lines for illustrative purposes.

SAWTOOTH WAVES

Besides the sine wave and the square wave, the *sawtooth wave* is common. This type of wave gets its name from its appearance when graphed. Figure 2-3 illustrates one type of sawtooth wave. The positive-going slope, also called the *attack, leading edge,* or *rise,* occurs instantly. But the negative-going slope, also known as the *release, trailing edge,* or *decay,* is gradual. The period of the wave is the time interval between points at identical positions on two successive pulses.

Another form of sawtooth wave is the opposite, with a gradual rise and an instant decay. This type of wave is also called a *ramp* (Fig. 2-4), and is commonly used for scanning in television sets and oscilloscopes.

Sawtooth waves can have rise and decay slopes in an infinite number of different combinations. One often-encountered example is shown in Fig. 2-5. In this case, the rise and decay are of equal, nonzero duration. This is a *triangular wave.*

COMPLEX AND IRREGULAR WAVES

Figure 2-6 shows an example of a *complex wave.* There is a definite period, and therefore a specific and measurable frequency. But it's hard to give a name to the shape; it appears irregular. But the waveform, although it is

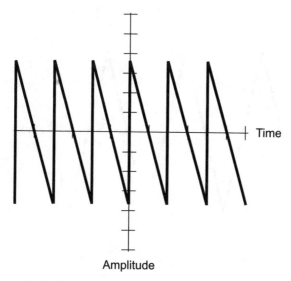

Fig. 2-3. A fast-rise, slow-decay sawtooth wave.

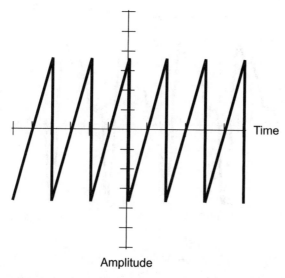

Fig. 2-4. A slow-rise, fast-decay sawtooth wave, also called a ramp wave.

not simple, nevertheless repeats itself. The period is the time between two points on adjacent, succeeding repetitions of the waveform.

An AC sine wave, as displayed on a lab instrument called a *spectrum analyzer*, appears as a single pip as shown in Fig. 2-7A. (The spectrum analyzer shows frequency on the horizontal scale and the amplitude on the

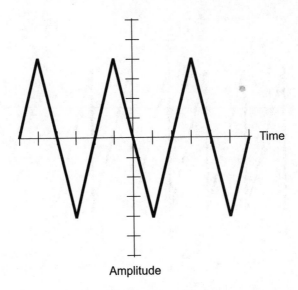

Time

Amplitude

Fig. 2-5. A triangular wave.

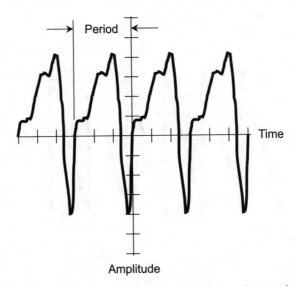

Period

Time

Amplitude

Fig. 2-6. A complex waveform. Four complete cycles are shown.

vertical scale, and depicts amplitude as a function of frequency. For this reason, it is called a *frequency-domain* display.) Some AC waves contain energy at harmonics, as well as at the fundamental frequency. If a wave has a frequency equal to *n* times the fundamental, then that wave is called the n*th harmonic*. In Fig. 2-7B, a wave is shown along with several harmonics, as it would look on the display of a spectrum analyzer.

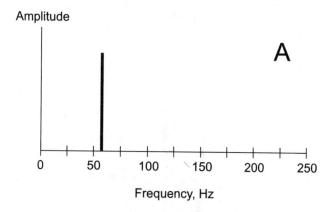

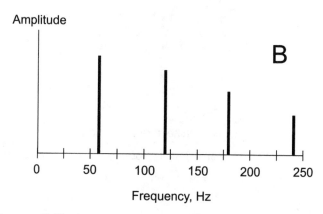

Fig. 2-7. At A, pure 60-Hz sine wave as seen on a spectrum-analyzer display. At B, a 60-Hz wave containing harmonic energy.

The frequency-domain displays of square waves and sawtooth waves show harmonic components in addition to the energy at the fundamental frequency. The wave shape depends on the amount of energy in the harmonics, and the way in which this energy is distributed among the harmonic frequencies.

Irregular waves can produce all sorts of weird frequency-domain displays. An example is shown at Fig. 2-8. This is a spectral display of an amplitude-modulated (AM) voice radio signal. Much of the energy is concentrated at the center of the pattern, at the frequency shown by the vertical line. There is also plenty of energy near, but not exactly at, the center frequency. (For now, don't worry about what the "decibels relative to 1 milliwatt" label on the vertical axis means. It's just a way engineers represent signal amplitude.)

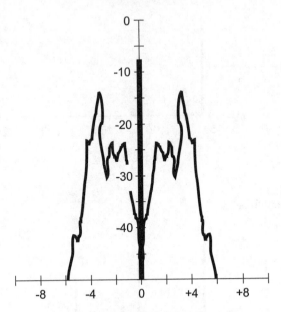

Fig. 2-8. An AM radio signal as it might look on a spectrum-analyzer display.

PROBLEM 2-1

The microsecond, symbolized μs, is one-millionth (0.000001 or 10^{-6}) of a second. What is the frequency, in hertz and in kilohertz, of a wave whose period is 50 μs?

SOLUTION 2-1

The frequency f, in hertz, is equal to the reciprocal of the period T, in seconds. In this case, $T = 0.000050$ seconds. Therefore:

$$f = 1/0.000050$$
$$= 20,000 \, \text{Hz}$$

This frequency is equivalent to 20 kHz.

PROBLEM 2-2

What is the frequency of the second harmonic of the wave in the preceding problem? The frequency of the third harmonic? The frequency of the nth harmonic, where n is any positive whole number?

SOLUTION 2-2

The fundamental frequency of this wave is 20 kHz. Let's call the frequency of the second harmonic (in kilohertz) f_2, the frequency of the third harmonic (in kilohertz) f_3, and in general, the frequency of the nth harmonic (in kilohertz) f_n. Then:

$$f_2 = 2 \times 20 = 40 \, \text{kHz}$$
$$f_3 = 3 \times 20 = 60 \, \text{kHz}$$
$$f_n = n \times 20 = 20n \, \text{kHz}$$

More Definitions

Engineers and scientists break the AC cycle down into small parts for analysis and reference. One popular method of describing an AC cycle is to divide it into 360 equal *degrees of phase*. The value 0° is assigned to the point in the cycle where the amplitude is 0 and positive-going. The same point on the next cycle is given the value 360°. The point halfway through the cycle is at 180°; a quarter cycle is 90°. The other method of specifying phase is to divide the cycle into 2π (approximately 6.2832) equal parts called *radians of phase*. A radian is equal to approximately 57.296°. This unit of phase is common among physicists.

ANGULAR FREQUENCY

Sometimes, the frequency of an AC wave is expressed in *degrees per second* rather than in hertz, kilohertz, megahertz, or gigahertz. Because there are 360° in one cycle, the *angular frequency* of a wave, in degrees per second, is 360 times the frequency in hertz. Angular frequency can also be expressed in *radians per second*. There are 2π radians in a complete cycle of 360°, so the angular frequency of a wave, in radians per second, is equal to 2π times the frequency in hertz.

AMPLITUDE EXPRESSIONS

The amplitude of an AC wave can be expressed in amperes (for current), volts (for voltage), or watts (for power).

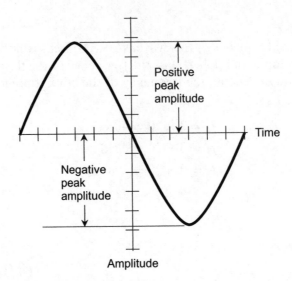

Fig. 2-9.   Positive and negative peak amplitudes. In this case they are equal.

The *instantaneous amplitude* is the amplitude at some precise moment in time. In an AC wave, the instantaneous amplitude varies as the wave cycle progresses. The manner in which it varies depends on the waveform. Instantaneous amplitude values are represented by individual points on the wave curves.

The *peak (pk) amplitude* is the maximum extent, either positive or negative, that the instantaneous amplitude attains. In some AC waves, the positive and negative peak amplitudes are the same, but sometimes they differ. Figure 2-9 shows a wave in which the positive peak amplitude is the same as the negative peak amplitude. Figure 2-10 is an illustration of a wave that has different positive and negative peak amplitudes.

The *peak-to-peak (pk-pk) amplitude* of a wave is the total difference between the positive peak amplitude and the negative peak amplitude (Fig. 2-11). Another way of saying this is that the peak-to-peak amplitude is equal to the *positive peak amplitude* plus the *negative peak amplitude*.

HYBRID AC/DC

An AC wave can have a *DC component* superimposed on it. If the *absolute value* of the DC component exceeds the absolute value of the positive or negative peak amplitude (whichever is larger) of the AC wave, then *pulsating DC* is the result. The absolute value of an electrical quantity is defined in the same way as absolute value of a quantity in basic algebra. The absolute value

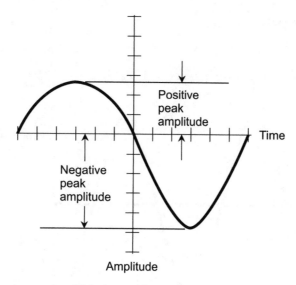

Fig. 2-10. A wave in which the positive and negative peak amplitudes differ.

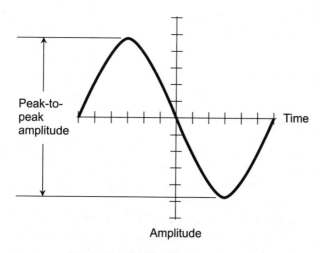

Fig. 2-11. Peak-to-peak amplitude.

of x is symbolized $|x|$, and represents the "distance of x from zero." If $x \geq 0$, then $|x| = x$. If $x < 0$, then $|x| = -x$.

Imagine a 200 V DC source connected in series with the 117 V AC from a standard utility outlet. Pulsating DC appears at the output with an average value of 200 V, but with instantaneous values much higher and lower. The waveform in this situation is illustrated by Fig. 2-12. The DC component is shown by the straight, dashed line, and the AC component is shown by the sine wave.

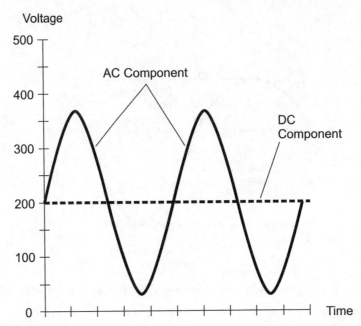

Fig. 2-12.   Composite AC/DC wave resulting from 117 volts rms AC in series with +200 volts DC.

ROOT-MEAN-SQUARE

Sometimes it is necessary to express the *effective amplitude* of an AC wave. This is the voltage, current, or power that a DC source would be required to produce for the same general effect. The most common specification is the *root-mean-square (rms) amplitude*.

For a perfect sine wave without any DC component, the rms amplitude is equal to approximately 0.7071 times the peak amplitude, or 0.3536 times the peak-to-peak amplitude. For a perfect square wave with no DC component, the rms amplitude is the same as the peak amplitude. For sawtooth and irregular waves with no DC component, the relationship between the rms and peak amplitudes depends on the shape of the waveform (the *waveform function*); this function must be expressed in mathematical terms for a precise calculation of the rms amplitude to be possible.

The existence of superimposed DC on a wave affects the rms amplitude. Calculating the rms value in such cases can be complicated. In any case, the rms amplitude can never be greater than the absolute value of the positive or negative peak amplitudes, whichever is larger.

PROBLEM 2-3

Suppose a sine wave has a peak-to-peak amplitude of 200 V, and consists of a pure AC wave with +50 V DC superimposed. What is the positive peak voltage? The negative peak voltage?

SOLUTION 2-3

The peak voltages are +100 V and −100 V with respect to the DC component. Because this component is +50 V, the positive and negative peak voltages (let's call them E_{pk+} and E_{pk-}) are:

$$E_{pk+} = (+50\,V) + (+100\,V) = +150\,V$$
$$E_{pk-} = (+50\,V) + (-100\,V) = -50\,V$$

PROBLEM 2-4

How much positive DC voltage must be superimposed on a wave whose positive and negative peak voltages are +150 V and −50 V, respectively, in order to produce a DC voltage as a result? How much negative voltage must be superimposed to produce a DC voltage as a result?

SOLUTION 2-4

It takes +50 V DC or −150 V DC to prevent the polarity from reversing. The DC voltage of +50 V brings the negative peak voltage up to zero, and the DC voltage of −150 V brings the positive peak voltage down to zero. In either of these situations, the polarity never reverses, so we have DC voltages. However, these DC voltages are pulsating, not steady.

Phase Relationships

When two sine waves have the same frequency, they can behave differently if their cycles begin at different times. Whether or not the *phase difference*, often called the *phase angle* and usually specified in degrees, is significant depends on the nature of the circuit.

PHASE COINCIDENCE

The phase relationship between two waves can have meaning only when the two waves have the same frequency. If the frequencies differ, the relative phase constantly changes. In the following discussions of phase angle, let's assume that the two waves always have the same frequency.

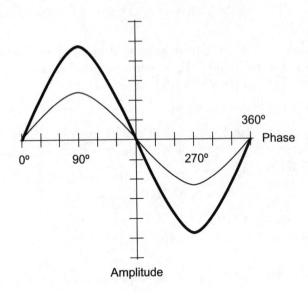

Fig. 2-13.   Two sine waves in phase coincidence.

Phase coincidence means that two waves begin at exactly the same instant in time. This is shown in Fig. 2-13 for two waves having different amplitudes. The phase difference in this case is 0°. If two sine waves are in phase coincidence, the peak amplitude of the resultant wave, which is also a sine wave, is equal to the sum of the peak amplitudes of the two composite waves. The phase of the resultant is the same as that of the composite waves.

PHASE OPPOSITION

When two waves begin exactly 1/2 cycle, or 180°, apart, they are said to be in *phase opposition*. This is illustrated by the drawing of Fig. 2-14. In this situation, engineers sometimes say that the waves are 180° out of phase.

If two sine waves have the same amplitude and are in phase opposition, they completely cancel each other, because the instantaneous amplitudes of the two waves are equal and opposite at every moment in time. If two sine waves have different amplitudes and are in phase opposition, the peak value of the resultant, which is a sine wave, is equal to the difference between the peak amplitudes of the two composite waves. The phase of the resultant is the same as the phase of the stronger of the two composite waves.

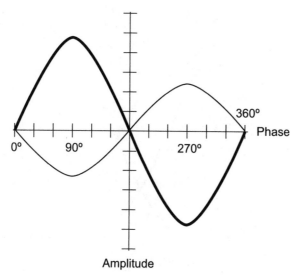

Fig. 2-14.   Two sine waves in phase opposition.

A sine wave has the unique property that, if its phase is shifted by 180°, the resultant wave is the same as turning the original wave "upside-down." Not all waveforms have this property.

LEADING PHASE

Two waves can differ in phase by any amount from 0° (in phase) through 180° (phase opposition) to 360° (in phase again).

Suppose there are two sine waves, X and Y, with identical frequencies. If wave X begins a fraction of a cycle earlier than wave Y, then wave X is said to be leading wave Y in phase. For this to be true, X must begin its cycle less than 180° before Y. Figure 2-15 shows wave X leading wave Y by 90°.

Note that if wave X (the thinner curve) is leading wave Y (the thicker curve), then wave X is somewhat to the left of wave Y in a graphical display. This is because, in a horizontal time-line display, the left-hand direction represents the past and the right-hand direction represents the future.

LAGGING PHASE

Suppose wave X begins its cycle more than 180°, but less than 360°, ahead of wave Y. In this situation, it is easier to imagine that wave X starts its cycle later than wave Y, by some value between 0° and 180°. Then wave X is

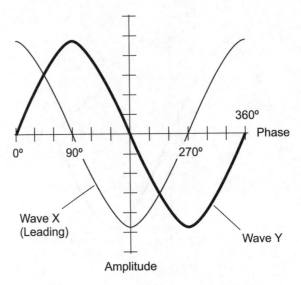

Fig. 2-15.   Wave X leads wave Y by 90°.

not leading, but instead is lagging, wave Y. Figure 2-16 shows wave X lagging wave Y by 90°.

If two waves have the same frequency and different phase, how do we know that one wave is leading the other by part of a cycle, instead of lagging by a cycle and a fraction, or by a few hundred, thousand, or million cycles and a fraction? The answer lies in the real-life effects. By convention, phase differences are expressed as values between 0° and 180°, either lagging or leading. Sometimes this range is given as −180° to +180° (lagging to leading).

Note that if wave X (the thinner curve in Fig. 2-16) is lagging wave Y (the thicker curve), then wave X is displaced somewhat to the right of wave Y on the graph.

VECTOR DIAGRAMS

Vectors are quantities that have two independent characteristics: *magnitude* and *direction*. They are illustrated as arrows in diagrams, with the length denoting the magnitude and the orientation representing the direction. They are denoted by bold type in text. Vector diagrams can be useful in illustrating phase relationships between sine waves that have the same frequency but not necessarily the same amplitude.

If a sine wave X is leading a sine wave Y by $q°$ (where q is some number between 0 and 180), then the two waves can be drawn as vectors, with

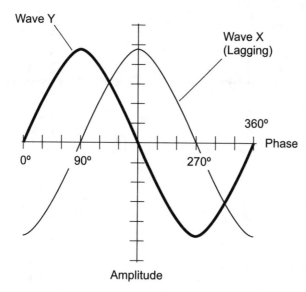

Fig. 2-16.   Wave X lags wave Y by 90°.

vector **X** being $q°$ of arc counterclockwise from vector **Y**. If wave X lags Y by $q°$ (where q, again, is some number between 0 and 180), then vector **X** is clockwise from vector **Y** by $q°$ of arc. If two waves are in phase, their vectors overlap (line up). If they are in phase opposition, they point in exactly opposite directions.

The drawings of Fig. 2-17 show four phase relationships between waves X and Y. Wave X always has twice the amplitude of wave Y, so vector **X** is always twice as long as vector **Y**. At A, wave X is in phase with wave Y. At B, wave X leads wave Y by 90°. At C, waves X and Y are opposite in phase. At D, wave X lags wave Y by 90°.

PROBLEM 2-5

Suppose two sine waves, wave X and wave Y, have identical frequencies and are in phase coincidence. Both waves have peak-to-peak voltages of 20 V, with no DC component. What are the positive peak, negative peak, and peak-to-peak voltages of the composite wave?

SOLUTION 2-5

The composite wave has twice the peak-to-peak amplitude of either wave alone, that is, $20 \times 2 = 40$ V.

Because there is no DC component in either wave, both of them have positive peak amplitudes of +10 V and negative peak amplitudes of −10 V. The positive peak amplitude of the composite is twice the positive peak

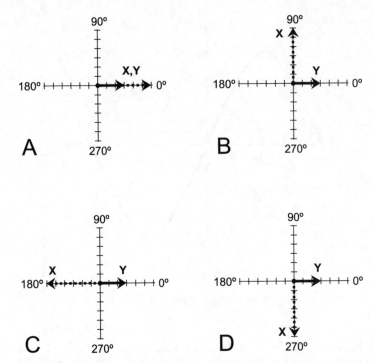

Fig. 2-17. At A, waves X and Y are in phase, so vectors X and Y point in the same direction. At B, wave X leads wave Y by 90°, so vector X is 90° counterclockwise from vector Y. At C, waves X and Y are in phase opposition, so vectors X and Y point in opposite directions. At D, wave X lags wave Y by 90°, so vector X is 90° clockwise from vector Y.

amplitude of either wave alone, that is, $+10 \times 2 = +20$ V. The same is true of the negative peak voltage of the composite; it is $-10 \times 2 = -20$ V.

PROBLEM 2-6
Suppose two sine waves, wave X and wave Y, have identical frequencies and are in phase opposition. Both waves have peak-to-peak voltages of 20 V, with no DC component. What are the positive peak, negative peak, and peak-to-peak voltages of the composite wave?

SOLUTION 2-6
The two waves cancel each other completely. Therefore, the positive peak, negative peak, and peak-to-peak voltages are all zero volts (0 V).

PROBLEM 2-7
Suppose two sine waves, wave X and wave Y, have identical frequencies and are in phase opposition. Wave X has a peak-to-peak voltage of 20 V, and wave Y has a peak-to-peak voltage of 12 V. Neither wave has a DC

component. What are the positive peak, negative peak, and peak-to-peak voltages of the composite wave?

SOLUTION 2-7
The composite wave has a peak-to-peak amplitude equal to the difference between the peak-to-peak voltages of waves X and Y. Peak-to-peak values are always expressed as positive numbers by convention, so the peak-to-peak voltage of the composite is derived by subtracting the smaller value from the larger. The peak-to-peak voltage of the composite wave is therefore $20 - 12 = 8$ V.

Because there is no DC component in wave X, it has a positive peak of $+10$ V. Wave Y also lacks a DC component, so negative peak is -6 V. The positive peak of wave X occurs at the same instant as the negative peak of wave Y. The values subtract because the waves are in phase opposition; the positive peak voltage of the composite is therefore $+10 - 6 = +4$ V.

Because there is no DC component in wave X, it has a negative peak of -10 V. Wave Y also lacks a DC component, so positive peak is $+6$ V. The negative peak of wave X occurs at the same instant as the positive peak of wave Y. The values subtract because the waves are in phase opposition; the negative peak voltage of the composite is therefore $-10 + 6 = -4$ V.

Utility Power Transmission

Electrical energy undergoes numerous transformations as it makes its way from the point of origin, usually an *electric generator*, to the end users. The initial source consists of potential or kinetic energy in some non-electrical form, such as falling or flowing water (*hydroelectric energy*), coal or oil (*fossil-fuel energy*), radioactive substances (*nuclear energy*), moving air (*wind energy*), light or heat from the sun (*solar energy*), or heat from the earth's interior (*geothermal energy*).

GENERATING PLANTS

In fossil-fuel, nuclear, geothermal, and some solar energy generating systems, heat is used to boil water, producing steam under pressure to drive *turbines*. These turbines produce the rotational force (torque) necessary to drive large *electric generators*. The greater the power demand becomes, the more torque is required to turn the generator shaft at the required speed.

In a *photovoltaic (PV) energy-generating system*, sunlight is converted into DC electricity directly by specialized semiconductor diodes called *photovoltaic cells*. This DC must be changed to AC in order to be used by most appliances. The conversion is done by a *power inverter*. Electrochemical batteries are necessary if a *stand-alone PV system* is to provide useful energy at night. Photovoltaic systems cannot generate much power, and are therefore used mainly in homes and small businesses. Stand-alone PV systems produce no waste products other than chemicals that must be discarded when storage batteries wear out. A PV system can be used in conjunction with existing utilities, without storage batteries, to supplement the total available energy supply to all consumers in a power grid. This is an *interactive PV system*. An interactive PV system for home use causes no environmental pollution of any kind.

In a *hydroelectric power plant*, the movement of water (waterfalls, tides, river currents) drives turbines that turn the generator shafts. In a *wind-driven electric power plant*, moving air operates devices similar to windmills, producing torque that turns the generator shafts. These power plants do not pollute directly, but large dams can disrupt ecosystems, have adverse affects on agricultural and economic interests downstream, and displace people upriver by flooding their land. Some people regard arrays of windmill-like structures as eyesores.

Whatever source is used to generate the electricity at a large power plant, the output is AC on the order of several hundred kilovolts, and in some cases more than a megavolt.

HIGH-TENSION LINES

When electric power is transmitted over wires for long distances, power is lost because of *ohmic resistance* in the conductors. This loss can be minimized in two ways. First, efforts can be made to keep the wire resistance to a minimum. This involves using large-diameter wires made from metal having excellent conductivity, and by routing the power lines in such a way as to keep their lengths as short as possible. Second, the highest possible voltage can be used. Long-distance, high-voltage power lines are called *high-tension lines*.

The reason that high voltage minimizes power-transmission line loss can be explained by looking at the equations denoting the relationship among power, current, voltage, and resistance. One equation can be stated as

$$P_{\text{loss}} = I^2 R$$

where P_{loss} is the power (in watts) lost in the line as heat, I is the line current (in amperes), and R is the line resistance (in ohms). For a given span of line, the value of R is a constant. The value of P_{loss} thus depends on the current in the line. This current depends on the *load*, that is, on the collective power demand of the end users.

Power-transmission line current also depends on line voltage. Current is inversely proportional to voltage at a given fixed power load. That is,

$$I = P_{\text{load}}/E$$

where I is the line current (in amperes), P_{load} is the total power demanded by the end users (in watts), and E is the line voltage (in volts). By making a substitution from this equation into the previous one, the following formula is obtained:

$$P_{\text{loss}} = (P_{\text{load}}/E)^2 R$$
$$= (P_{\text{load}})^2 R/E^2$$

For any given fixed values of P_{load} and R, doubling the line voltage E reduces the power loss P_{loss} to one-quarter (25%) of its previous value. If the line voltage is multiplied by 10, the theoretical line power loss drops to 1/100 (1%) of its former value. Therefore, it makes good technological sense to generate as high a voltage as possible for efficient long-distance electric power transmission.

The loss in a power line can be reduced further by using DC, rather than AC, for long-distance transmission. Direct currents do not produce *electromagnetic (EM) fields* as alternating currents do, so a secondary source of power-line inefficiency, known as *EM radiation loss*, is eliminated by the use of DC. However, long-distance DC power transmission is difficult and costly to implement. It necessitates high-power, high-voltage rectifiers at a generating plant, and power inverters at distribution stations where high-tension lines branch out into lower-voltage, local lines.

TRANSFORMERS

The standard utility voltages are 117 V and 234 V rms. These comparatively low voltages are obtained from higher voltages by using *transformers*.

Step-down transformers reduce the voltage of high-tension lines (100,000 V or more) down to a few thousand volts for distribution within municipalities. These transformers are physically large because they must carry significant power. Several of them can be placed in a building or a fenced-off area.

The outputs of these transformers are fed to power lines that run along city streets.

Smaller transformers, usually mounted on utility poles or underground, step the municipal voltage down to 234 V rms for distribution to individual homes and businesses. The 234 V electricity is provided in the form of three separate AC waves, called *phases*, to the distribution boxes in each house, apartment, or business. Each wave is 120° out of phase with the other two. Some utility outlets are supplied directly with such *three-phase AC*. Large appliances such as electric stoves, ovens, and laundry machines use electricity in this form. Conventional wall outlets in the United States are provided with single-phase, 117 V rms AC.

PROBLEM 2-8

Suppose a span of high-tension utility line has an effective resistance of $100\,\Omega$, and the effective voltage on the line is 500 kV. What is the power loss?

SOLUTION 2-8

We can't answer this unless we know the power demand at the load, or the current carried by the line. If there is no power demand at the load end of the line, then there is no current in the line, and therefore no loss in it. But if there is a substantial power demand at the end of the line, resulting in a large current, the power loss is considerable. As the demand increases, so does the line loss.

PROBLEM 2-9

Suppose that in the above situation, the consumers at the end of the high-tension line collectively demand 500 kilowatts (500 kW or 500,000 W) of power. What is the line loss?

SOLUTION 2-9

Use the formula for the power loss P_{loss} in terms of the load power P_{load}, the line resistance R, and the line voltage E. In this instance:

$$P_{load} = 500,000\,\text{W}$$
$$R = 100\,\Omega$$
$$E = 500,000\,\text{V}$$

The value of P_{loss} can be calculated easily. The answer might surprise you:

$$P_{loss} = (P_{load})^2 R/E^2$$
$$= 500,000^2 \times 100/500,000^2$$
$$= 100\,\text{W}$$

This is an efficient power transmission line! Think of it this way: In order to deliver half a million watts to customers, the power generator need only produce approximately 100 extra watts to make up for the heat loss in the wires. Keep in mind that this is an idealized scenario, and the numbers were manufactured for this theoretical exercise. In real life, things rarely work out so well.

Quiz

Refer to the text in this chapter if necessary. Answers are in the back of the book.

1. A photovoltaic cell produces
    (a) direct current.
    (b) triangular-wave current.
    (c) square-wave current.
    (d) alternating current.

2. The AC voltage of utility electricity can be reduced by
    (a) a transformer.
    (b) a PV cell.
    (c) a high-tension line.
    (d) an electric generator.

3. Refer to Fig. 2-18. On the horizontal scale, each division represents 1 millisecond (1 ms), which is 1/1000 of a second (0.001 s). On the vertical scale, each division represents 10 V. As you go upward on the vertical scale, the voltage becomes more positive, and as you go downward, the voltage becomes more negative. At the origin, the voltage is zero. What is the approximate frequency of this wave?
    (a) 333 Hz
    (b) 556 Hz
    (c) 833 Hz
    (d) It cannot be determined without knowing the waveform function.

4. What is the approximate period of the wave shown in Fig. 2-18?
    (a) 1.2 ms.
    (b) 1.8 ms.
    (c) 3 ms.
    (d) It cannot be determined without knowing the waveform function.

Each horizontal division = 1 ms
Each vertical division = 10 V

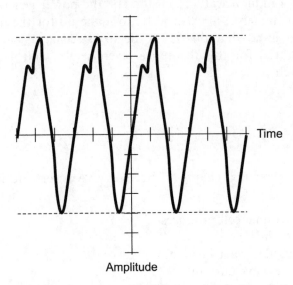

Time

Amplitude

Fig. 2-18.   Illustration for Quiz Questions 3 through 8.

5. What is the approximate positive peak voltage of the wave shown in Fig. 2-18?
    (a) +90 V
    (b) +50 V
    (c) +40 V
    (d) It cannot be determined without knowing the waveform function.

6. What is the approximate negative peak voltage of the wave shown in Fig. 2-18?
    (a) −90 V
    (b) −50 V
    (c) −40 V
    (d) It cannot be determined without knowing the waveform function.

7. What is the approximate peak-to-peak voltage of the wave shown in Fig. 2-18?
    (a) 90 V
    (b) 50 V
    (c) 40 V
    (d) It cannot be determined without knowing the waveform function.

8. What is the approximate rms voltage of the wave shown in Fig. 2-18?
    (a) 28.3 V
    (b) 31.9 V
    (c) 35.4 V
    (d) It cannot be determined without knowing the waveform function.

9. If two waves having identical frequencies are in phase opposition, their periods are
    (a) opposite.
    (b) the same.
    (c) different.
    (d) impossible to determine.

10. If all the energy in an AC wave is concentrated at a single frequency, and there is no harmonic energy whatsoever, then the waveform is
    (a) square.
    (b) sawtooth.
    (c) triangular.
    (d) sinusoidal.

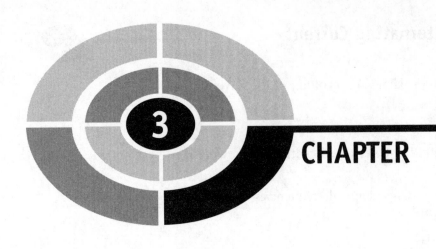

CHAPTER

3

Impedance

Impedance is the opposition that a component or circuit offers to AC. Impedance (symbolized by Z) consists of two independent components, *resistance* (symbolized by R) and *reactance* (symbolized by X). In order to understand impedance fully, we must know about *imaginary numbers* and *complex numbers*.

Introducing *j*

Have you ever wondered what the square root of a negative number is? You are about to find out, if you have not yet learned about imaginary numbers. Imaginary numbers are no more or less genuine than real numbers such as 2, −5, 83/2, or the cube root of 17. The imaginary numbers got their label because they're a little harder to directly envision than ordinary numbers.

The set of imaginary numbers derives from the positive square root of −1, called the *unit imaginary number i* by mathematicians and the *j operator* by engineers. This is the number that, when multiplied by itself, results in a product of -1. We'll use the engineering notation and call it j. Thus:

$$j \times j = -1$$

It also happens to be true that $-j$, when multiplied by itself, equals -1. This is because the product of two negatives always produces a positive:

$$(-j) \times (-j) = -1$$

Two more tidbits can finish up this mathematical exercise:

$$(-j) \times j = 1$$

$$j \times (-j) = 1$$

The number j can be multiplied by any real number X, getting an imaginary number jX. The set of all possible imaginary numbers can be represented as an *imaginary number line* that is similar to the *real number line*. The imaginary number line can be placed at a right angle to the real number line and joined with it at the zero points to form the *complex number plane* (Fig. 3-1). This, it turns out, has useful properties that has made the work of engineers easier than it might otherwise have been.

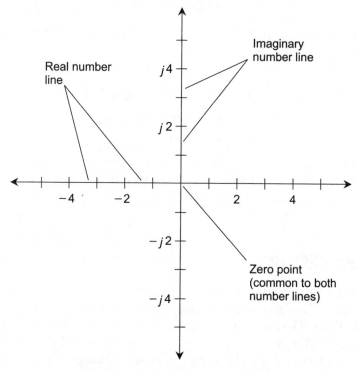

Fig. 3-1.   The complex number plane.

COMPLEX NUMBERS

When you add a real number R and an imaginary number jX, you get a *complex number*. (A better term might be *composite number*, but the term *complex* was used by the people who first worked with these numbers, and it has stuck.)

Real numbers are one-dimensional, expressible on a simple straight-line scale, and are therefore called *scalar quantities*. Imaginary numbers are also one-dimensional and can be represented on a simple scale, so they are scalar too. But complex numbers need two dimensions to be defined, and because of this, they are *vector quantities*. A complex number can be expressed in the form $R+jX$, where R and X are both real numbers. It can also be expressed as a point on a two-dimensional coordinate plane.

Adding complex numbers requires adding the real parts and the complex parts separately. For example, the sum of $4+j7$ and $45-j83$ is found this way:

$$(4+45)+j(7-83)$$
$$=49+j(-76)$$
$$=49-j76$$

Subtracting complex numbers works similarly. The difference $(4+j7)-(45-j83)$ is found like this:

$$(4+j7)-(45-j83)$$
$$(4+j7)+[-1(45-j83)]$$
$$=(4+j7)+(-45+j83)$$
$$=-41+j90$$

Subtraction can always be performed as an addition of a negative quantity, as is done in the second line of the above calculation. Remember this, and you minimize the risk of getting signs confused within mathematical expressions.

COMPLEX VECTORS

Any complex number $R+jX$ can be represented as a vector in the complex number plane. This gives each complex number a unique *magnitude* and a unique *direction*. The magnitude is the distance of the point (R,X), representing the number $R+jX$, from the origin $(0,0)$, representing the number $0+j0$. The direction is the angle the vector subtends, measured counterclockwise from the $+R$ axis as shown in Fig. 3-2. (Don't think of R as resistance

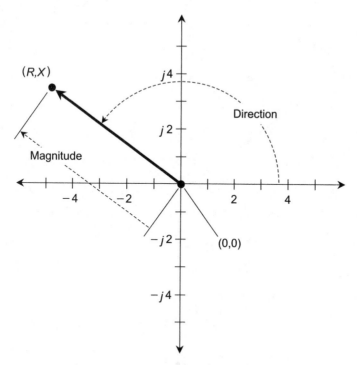

Fig. 3-2.   Magnitude and direction of a vector in the complex number plane.

here, but merely as the real-number component of a complex number. Then you won't be puzzled by what looks like a negative resistance in this example.)

The *absolute value* of a complex number $R+jX$, denoted $|R+jX|$, is the length of its vector (R, X) in the complex plane, measured from the origin $(0,0)$ to the point (R, X). In the case of a pure positive real number or 0, the absolute value is simply the number itself. In the case of a negative real number, the absolute value is the number multiplied by -1. In the case of a pure imaginary number $0+jX$, the absolute value is equal to X when X is positive, and X multiplied by -1 if X is negative.

If a complex number is neither pure real nor pure imaginary, its absolute value can be found using a formula based on the *Pythagorean theorem* from plane geometry. You might remember this as the formula used to find the length of the longest side (or *hypotenuse*) of a right triangle. First, square both R and X. Then add them. Finally, take the square root of the sum. This is Z, the length of the vector (R,X) as shown in Fig. 3-3. Mathematically:

$$Z = |R+jX| = (R^2 + X^2)^{1/2}$$

where the 1/2 power represents the square root.

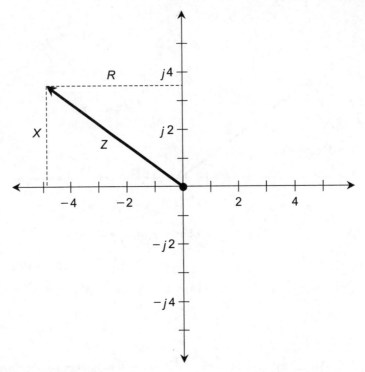

Fig. 3-3. The absolute value of a complex number is the length of its vector in the complex plane. In this example, the distances R, X, and Z are related by the equation $Z^2 = R^2 + X^2$. Therefore, $Z = (R^2 + X^2)^{1/2}$.

Inductive Reactance

Resistance in DC circuits is a scalar quantity. Given a constant DC voltage, the current decreases as the resistance increases, in accordance with Ohm's Law. The same law holds for AC through a pure resistance. But in a component that has *inductance* (such as a coil of wire) or capacitance (such as a pair of large metal plates placed close together but not touching), the AC situation is more complicated.

COILS AND CURRENT

If you wind a length of wire into a coil (thereby making it into an *inductor*) and connect it to a source of DC, the coiled-up wire draws an electrical current I that is equal to E/R, where I is the current in amperes, E is the

DC voltage in volts, and R is the DC resistance of the wire in ohms. The coil gets hot as energy is dissipated in the resistance of the wire. If the voltage of the power supply is increased, the wire gets hotter. Eventually, if connected to a substantial enough power supply for a long enough period of time, the wire will melt.

Suppose you change the voltage across the coil from DC to AC. You vary the frequency from a few hertz (Hz) to many megahertz (MHz). The coil has a certain *inductive reactance* (denoted X_L), which is a sort of "electrical sluggishness." Because of this "sluggishness," it takes some time for current to establish itself in the coil. Inductive reactance is expressed in ohms, just like resistance. The standard unit of inductance is called the *henry*, abbreviated by a non-italicized, uppercase letter H.

As the AC frequency keeps increasing, a point is reached beyond which the current can't get well established in the coil before the polarity of the voltage reverses. As the frequency is raised further still, this effect becomes more and more pronounced. Eventually, if the frequency becomes high enough, the coil cannot come anywhere near establishing a significant current with each cycle. Therefore, almost no current flows through the coil. The coil remains cool even if the power supply is capable of delivering a lot of power, which would melt the coil in a hurry if it were in the form of DC.

The X_L of a coil or inductor can vary from zero to many megohms. Like pure resistance, inductive reactance affects the current in an AC circuit. But unlike pure resistance, reactance changes with frequency.

X_L VERSUS FREQUENCY

If the frequency of an AC source is given (in hertz) as f, and the inductance of a coil is given (in henrys) as L, then the inductive reactance (in ohms), X_L, is given by:

$$X_L = 2\pi f L$$

where π is equal to approximately 3.14159. This same formula applies if f is in kilohertz and L is in *millihenrys* (mH), where $1\,\text{mH} = 0.001\,\text{H}$. It also applies if f is in megahertz and L is in *microhenrys* (μH), where $1\,\mu\text{H} = 0.000001\,\text{H}$.

Inductive reactance increases with increasing AC frequency. Inductive reactance also increases with inductance, when the frequency is held constant. The value of X_L is directly proportional to f and also to L. These relationships are graphed, in relative form, in Fig. 3-4.

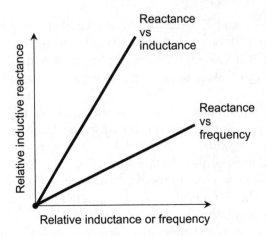

Fig. 3-4.   Inductive reactance is directly proportional to inductance (L) and also to frequency (f).

POINTS IN THE RX_L PLANE

Inductive reactance can be plotted along a half-line. So can resistance. In a circuit containing both resistance (R) and inductive reactance (X_L), the characteristics become two-dimensional. You can orient the resistance and reactance half-lines perpendicular to each other to define a coordinate plane. In this scheme, RX_L combinations form *complex impedances*. Each point on the plane corresponds to a unique impedance, and each RX_L impedance value corresponds to a unique point on the plane. Impedances on this RX_L *plane* are written in the form $R + jX_L$, where R is the resistance (represented by a non-negative real number) and X_L is the inductive reactance (also represented by a non-negative real number).

If you have a resistance $R = 5\,\Omega$, the complex impedance is $5 + j0$, and is represented by the point (5,0) on the plane. If you have a pure inductive reactance, such as $X_L = 3\,\Omega$, the complex impedance is $0 + j3$, and is at the point (0,3) on the plane. Often, resistance and inductive reactance exist together. This gives impedance values such as $2 + j3$ or $4 + j1.5$. Figure 3-5 shows some points on the RX_L plane, each of which represents a specific impedance.

VECTORS IN THE RX_L PLANE

Engineers represent points in the RX_L plane as vectors. This gives each point a specific and unique magnitude and direction.

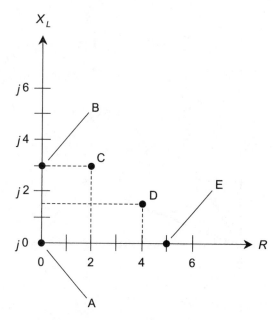

Fig. 3-5. Some points in the RX_L impedance plane. At A, $0+j0$. At B, $0+j3$. At C, $2+j3$. At D, $4+j1.5$. At E, $5+j0$.

Another way to think of the points shown in Fig. 3-5 is to draw lines from the origin out to them. Then the points become rays, each having a certain magnitude and a direction (angle counterclockwise from the R axis). These rays, going out to the points, are *complex impedance vectors* (Fig. 3-6).

Current and Voltage in *RL* Circuits

An electronic component containing inductance stores electrical energy as a *magnetic field*. When a voltage is placed across a coil, it takes a while for the current to build up to full value. Therefore, when AC is placed across a coil, the current lags behind the voltage in phase.

INDUCTANCE AND RESISTANCE

Imagine that you connect a source of AC voltage across a low-resistance coil. Suppose the AC frequency is high enough so X_L is much larger than the DC resistance, R. In this situation, the current lags the voltage by almost

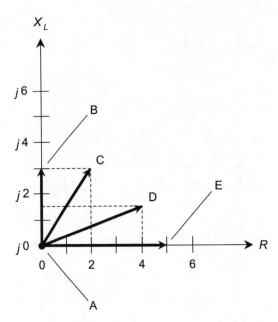

Fig. 3-6. Some vectors in the RX_L impedance plane. At A, $0+j0$. At B, $0+j3$. At C, $2+j3$. At D, $4+j1.5$. At E, $5+j0$.

$90°$ (1/4 cycle). When the value of X_L is gigantic compared with the value of R, the RX_L-plane vector points nearly along the X_L axis. Its angle is almost $90°$ from the R axis.

When the resistance in a *resistance-inductance (RL) circuit* is significant compared with the reactance, the current lags the voltage by less than $90°$ (Fig. 3-7). If R is small compared with X_L, the current lag is nearly $90°$; as R gets relatively larger, the lag decreases. When R is many times greater than X_L, the vector lies almost on the R axis, and the *RL phase angle* is only a little more than $0°$. When the complex impedance becomes a pure resistance, the current comes into phase with the voltage, and the vector lies exactly along the R axis. (Then we no longer have an *RL* circuit, but simply an R circuit!)

CALCULATING PHASE ANGLE IN *RL* CIRCUITS

You can use a ruler that has centimeter (cm) and millimeter (mm) markings, along with a protractor, to approximate the phase angle in an *RL* circuit. First, draw a horizontal line a little more than 100 mm long, going from left to right. Construct a vertical line off the left end of this first line, going vertically upwards. Make this line at least 100 mm long. The horizontal line is the R axis. The vertical line is the X_L axis (Fig. 3-8).

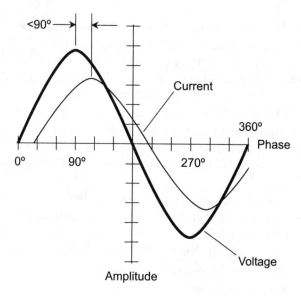

Fig. 3-7. In an RL circuit, the current lags the voltage by some phase angle greater than $0°$ but less than $90°$.

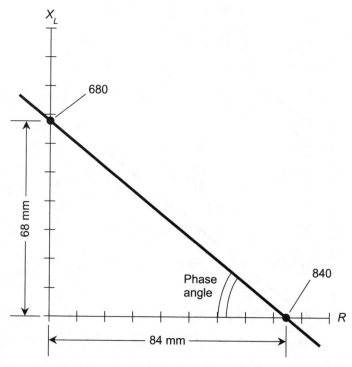

Fig. 3-8. Pictorial method of finding the phase angle in an RL circuit.

If you know the values of X_L and R, divide them down or multiply them up so they are both between 0 and 100. For example, if $X_L = 680\,\Omega$ and $R = 840\,\Omega$, divide them both by 10 to get $X_L = 68$ and $R = 84$. Plot these points by making hash marks on the axes. In this example, the R mark should be 84 mm to the right of the origin, and the X_L mark should be 68 mm up from the origin. Draw a line connecting the hash marks, as shown in Fig. 3-8. This line forms, along with the two axes, forms a right triangle. Measure the angle between the slanted line and the R axis. This is the RL phase angle, symbolized ϕ_{RL}. (The symbol ϕ is the lowercase Greek letter phi, pronounced "fie" or "fee.")

The actual vector corresponding to a complex impedance $R + jX_L$ is found by constructing a rectangle using the origin and hash marks as three of the four vertices, and drawing horizontal and vertical lines to complete the figure. The vector is the diagonal of this rectangle, as shown in Fig. 3-9. The angle ϕ_{RL} is the angle between this vector and the R axis.

A more exact value for ϕ_{RL}, especially if it is near $0°$ or $90°$ making pictorial constructions difficult, can be found using a calculator that has

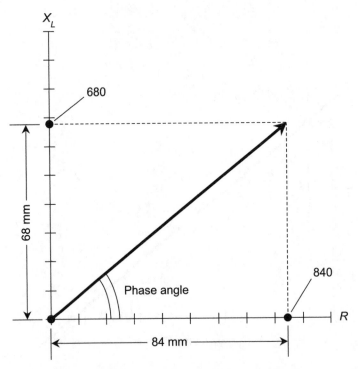

Fig. 3-9.   Illustration of an RX_L impedance vector.

inverse trigonometric functions. The angle is the arctangent (arctan or $\tan^{-1}$) of the ratio of inductive reactance to resistance:

$$\phi_{RL} = \arctan(X_L/R)$$

PROBLEM 3-1

A circuit contains 30 Ω of resistance and 40 Ω of inductive reactance. What is the complex impedance? What is the phase angle to the nearest degree?

SOLUTION 3-1

In this example, $R = 30\,\Omega$ and $X_L = 40\,\Omega$. The complex impedance Z is:

$$Z = R + jX_L$$
$$= 30 + j40$$

The phase angle ϕ_{RL} is:

$$\phi_{RL} = \arctan(X_L/R)$$
$$= \arctan(40/30)$$
$$= \arctan 1.333$$
$$= 53° \text{ (rounded to nearest degree)}$$

PROBLEM 3-2

Suppose the frequency in the above example is 1000 Hz. What inductance, to the nearest tenth of a millihenry, corresponds to a reactance of 40 Ω in this case?

SOLUTION 3-2

To solve this, "plug in" the known values to the equation for inductive reactance in terms of frequency and inductance:

$$X_L = 2\pi f L$$
$$40 = 2 \times 3.14159 \times 1000 \times L$$

We want to find L. This is done by dividing both sides of the equation by $2 \times 3.14159 \times 1000$, and then calculating it out:

$$L = (40)/(2 \times 3.14159 \times 1000)$$
$$= 40/6283.18$$
$$= 0.00637\,\text{H}$$
$$= 6.4\,\text{mH (rounded to nearest 0.1 mH)}$$

Capacitive Reactance

Inductive reactance has an "evil twin" in *capacitive reactance*. This, too, can be represented as a scalar quantity, starting at the same zero point as inductive reactance, but running off in the opposite direction, having negative ohmic values.

CAPACITORS AND CURRENT

Imagine two gigantic, flat, parallel metal plates, both of which are excellent electrical conductors. These plates form a *capacitor*. (In some old texts, you'll see the term *condenser* used instead.) If supplied with DC, the plates become electrically charged, and eventually they acquire a potential difference equal to the DC source voltage. The current, once the plates are charged, is zero.

Suppose the power supply is changed from DC to AC. Imagine that you can adjust the frequency of this AC from a few hertz to many megahertz. At first, the voltage between the plates follows almost exactly along as the AC polarity reverses; the set of plates acts like an open circuit. As the frequency increases, the charge cannot get well established with each cycle, so current keeps flowing into and out of the plates consistently as the voltage polarity goes back and forth. When the frequency becomes extremely high, the plates don't even come close to getting fully charged with each cycle; the current flows continuously, and the pair of plates behaves almost like a short circuit.

Capacitive reactance (symbolized X_C) quantifies the opposition that a capacitor offers to AC. Like inductive reactance, X_C varies with frequency and is measured in ohms. But X_C is, by convention, assigned negative ohmic values rather than positive ones. The value of X_C increases negatively as the frequency goes down. The unit of capacitance is called the *farad*, abbreviated by a non-italicized, uppercase letter F.

Sometimes X_C is talked about in terms of its absolute value, with the minus sign removed. But in complex impedance calculations, X_C is always negative or zero. This prevents it from getting confused with inductive reactance, X_L, which is always positive or zero.

X_C VERSUS FREQUENCY

In some ways, capacitive reactance behaves like a mirror image of inductive reactance. In another sense, X_C can be imagined as an extension, rather than a reflection, of X_L into negative values.

If the frequency of an AC source is given (in hertz) as f, and the value of a *capacitor* is given (in farads) as C, then the capacitive reactance (in ohms), X_C, can be calculated as follows:

$$X_C = -1/(2\pi f C)$$

This same formula applies if the frequency is in megahertz and the capacitance is in *microfarads* (μF), where $1\,\mu$F $= 0.000001$ F. Note that in the case of a capacitive reactance, we have "negative ohms," something that you never hear about with DC resistances.

For some reason, *millifarads* (mF) are almost never specified as a unit of capacitance. In fact, the appearance of the abbreviation mF can cause confusion. If you ever see a component labeled as something like 100 mF, the engineers usually mean to say it is 100 μF.

Capacitive reactance is negatively, and inversely, proportional to frequency. The function X_C vs f appears as a curve when graphed, and this curve "blows up" as the frequency nears zero. Capacitive reactance also varies inversely with the actual value of capacitance, given a fixed frequency. Therefore, the function of X_C vs C appears as a curve that "blows up" as the capacitance approaches zero. Relative graphs are shown in Fig. 3-10.

POINTS IN THE RX_C PLANE

In a circuit containing resistance and capacitive reactance, the characteristics are two-dimensional, in a way that is analogous to the situation with the RX_L plane. The resistance ray and the capacitive-reactance ray can be

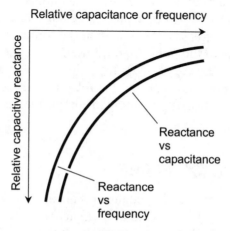

Fig. 3-10. Capacitive reactance is negatively, and inversely, proportional to frequency (f) and also to capacitance (C).

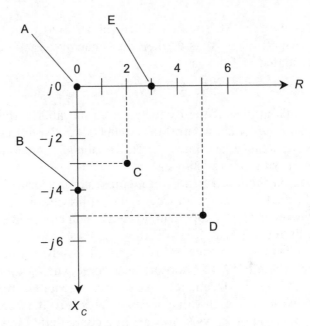

Fig. 3-11.   Some points in the RX_C impedance plane. At A, $0 - j0$. At B, $0 - j4$. At C, $2 - j3$. At D, $5 - j5$. At E, $3 - j0$.

placed end-to-end at right angles to make the RX_C *plane* (Fig. 3-11). Resistance is plotted horizontally, with increasing values toward the right. Capacitive reactance is plotted vertically, with increasingly negative values downward. Each point corresponds to exactly one complex impedance. Conversely, each specific impedance coincides with exactly one point on the plane. Impedance values that contain resistance and capacitance can be written in complex form. Because X_C is never positive, engineers usually write $R - jX_C$.

If the resistance is pure, say $R = 3\,\Omega$, then the complex impedance is $3 - j0$, and this corresponds to the point (3,0) on the RX_C plane. If you have a pure capacitive reactance, say $X_C = -4\,\Omega$, then the complex impedance is $0 - j4$, and this is at the point (0,−4) on the RX_C plane. Often, resistance and capacitive reactance both exist, resulting in complex impedances such as $2 - j3$ or $5 - j5$.

VECTORS IN THE RX_C PLANE

In Fig. 3-11, several complex impedance points are shown, corresponding to various combinations of resistance and capacitive reactance. Each point is represented by a certain distance to the right of the origin (0,0), and a certain displacement downwards.

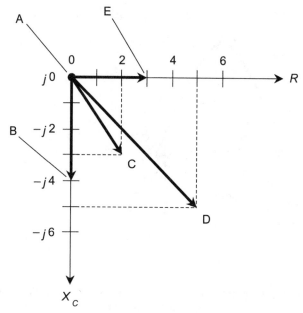

Fig. 3-12.   Some vectors in the RX_C impedance plane. At A, $0 - j0$. At B, $0 - j4$. At C, $2 - j3$. At D, $5 - j5$. At E, $3 - j0$.

Impedance points in the RX_C plane can be rendered as vectors, just as can be done in the RX_L plane. The magnitude of a vector is the distance of the point from the origin; the direction is the angle measured clockwise from the resistance (R) line, and specified in *negative degrees*. The vectors for the points of Fig. 3-11 are depicted in Fig. 3-12.

Current and Voltage in *RC* Circuits

A capacitor stores energy in the form of an *electric field*. When an AC voltage source is placed across a circuit containing resistance and capacitance, the current leads the voltage in phase. The phase difference is always more than 0°, and always less than a quarter of a cycle (90°).

CAPACITANCE AND RESISTANCE

When the resistance in a *resistance-capacitance (RC) circuit* is significant compared with the capacitive reactance, the current leads the voltage by something less than 90° (Fig. 3-13). If R is small compared with X_C,

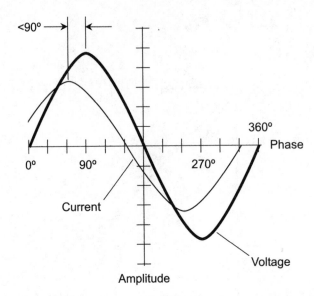

Fig. 3-13. In an *RC* circuit, the current leads the voltage by some phase angle greater than 90° but less than 90°.

the difference is almost 90°. As *R* gets larger, or as X_C becomes smaller, the phase difference decreases. The value of *R* can increase relative to X_C because resistance is deliberately put into a circuit, or because the frequency becomes so low that X_C rises to a value comparable with the *leakage resistance* of the capacitor. In either case, the situation can be represented by a resistor and capacitor in series.

As the resistance in an *RC* circuit gets large compared with the capacitive reactance, the phase angle gets closer and closer to 0°. The same thing happens if the value of X_C gets small compared with the value of *R*. (When we call X_C "large," we mean "large negatively." When we say X_C is "small," we mean that it is close to zero, or "small negatively.")

When *R* is many times larger than X_C, the vector in the RX_C plane lies almost along the *R* axis. Then the RX_C phase angle is close to 0°. Ultimately, if the capacitive reactance gets small enough, the circuit behaves as a pure resistance, and the current comes into phase with the voltage, and the impedance vector lies along the *R* axis.

CALCULATING PHASE ANGLE IN *RC* CIRCUITS

If you know the ratio X_C/R in an *RC* circuit, then you can find the phase angle. You can use a protractor and a ruler to find phase angles for *RC* circuits, just as with *RL* circuits, as long as the angles are not too close to 0° or 90°.

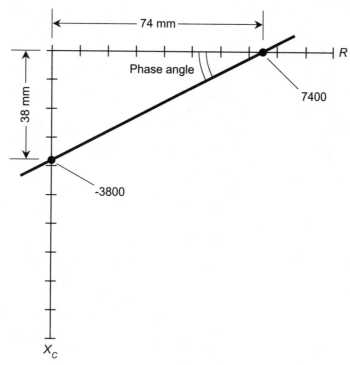

Fig. 3-14.   Pictorial method of finding phase the angle in an *RC* circuit.

First, draw a line somewhat longer than 100 mm, going from left to right on the paper. Then, use the protractor to construct a line going somewhat more than 100 mm vertically downwards, starting at the left end of the horizontal line. The horizontal line is the *R* axis of the RX_C plane. The line going down is the X_C axis (Fig. 3-14).

If you know the values of X_C and *R*, divide or multiply them by a constant, chosen to make both values fall between -100 and 100. For example, if $X_C = -3800 \, \Omega$ and $R = 7400 \, \Omega$, divide them both by 100, getting -38 and 74. Plot these points on the lines as hash marks. The X_C mark goes 38 mm down from the origin. The *R* mark goes 74 mm to the right of the origin. Draw a line connecting the hash marks, as shown in Fig. 3-14. Measure the angle between the slanted line and the *R* axis. This angle is between $0°$ and $90°$. Multiply by -1 to get the phase angle for the *RC* circuit, symbolized ϕ_{RC}.

The complex impedance vector is diagrammed by constructing a rectangle using the origin and the hash marks, making new perpendicular lines to complete the figure. The vector is the diagonal of this rectangle, running out from the origin (Fig. 3-15). The phase angle is the angle between the *R* axis and this vector, multiplied by -1.

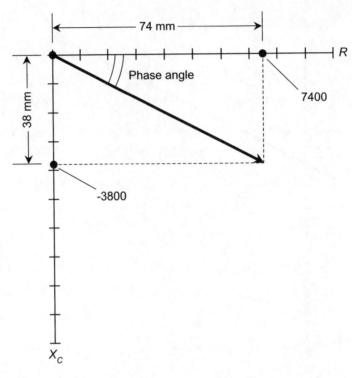

Fig. 3-15. Illustration of an RX_C impedance vector.

The more accurate way to find RC phase angles involves trigonometry. Determine the ratio X_C/R and enter it into a calculator. This ratio is a negative number or zero, because X_C is always negative or zero, and R is always positive. Find the arctangent (arctan or $\tan^{-1}$) of this number. This is ϕ_{RC}. Mathematically:

$$\phi_{RC} = \arctan(X_C/R)$$

PROBLEM 3-3

A circuit contains $100\,\Omega$ of resistance and $-100\,\Omega$ of capacitive reactance. What is the complex impedance? What is the phase angle to the nearest degree?

SOLUTION 3-3

In this example, $R = 100\,\Omega$ and $X_L = -100\,\Omega$. The complex impedance Z is:

$$Z = R + jX_C$$
$$= 100 - j100$$

The phase angle ϕ_{RC} is:

$$\phi_{RC} = \arctan(X_C/R)$$

$$= \arctan(-100/100)$$

$$= \arctan -1$$

$$= -45° \text{ (rounded to nearest degree)}$$

PROBLEM 3-4
Suppose the frequency in the above example is 1000 kHz. What capacitance, to the nearest ten-thousandth of a microfarad, corresponds to a reactance of $-100\,\Omega$ in this case?

SOLUTION 3-4
Note that $1000\,\text{kHz} = 1.000\,\text{MHz}$. First, "plug in" the known values to the equation for capacitive reactance in terms of frequency (in megahertz) and capacitance (in microfarads):

$$X_C = -1/(2\pi fC)$$

$$-100 = -1/(2 \times 3.14159 \times 1.000 \times C)$$

$$100 = 1/(2 \times 3.14159 \times 1.000 \times C)$$

$$0.01 = (2 \times 3.14159 \times 1.000 \times C)$$

We want to find C. This is done by dividing both sides of the equation by $2 \times 3.14159 \times 1.000$, and then calculating:

$$C = (0.01)/(2 \times 3.14159 \times 1.000)$$

$$= 0.01/6.28318$$

$$= 0.00159155\,\mu\text{F}$$

$$= 0.0016\,\mu\text{F} \text{ (rounded to nearest } 0.0001\,\mu\text{F})$$

THE *RX* PLANE

Recall the plane for R and X_L. This forms the upper portion of an *RX plane* shown in Fig. 3-16. Similarly, the plane for R and X_C forms the lower portion of the *RX* plane. Resistances are represented by non-negative real numbers. Reactances, whether inductive (positive) or capacitive (negative), correspond to imaginary numbers. Any specific impedance $R + jX$ can be represented by

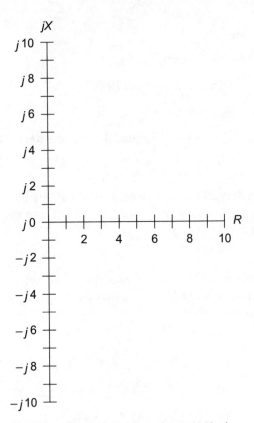

Fig. 3-16. The complex impedance (*RX*) plane.

a unique complex number, where *R* is a non-negative real number and *X* is any real number.

Sometimes, the italicized, uppercase letter *Z* is used in place of the word "impedance" in general discussions, and yet the figures given are real numbers, not complex numbers. Suppose a dynamic loudspeaker is advertised as having $Z = 8\ \Omega$. If no specific complex impedance is given, this can, in theory, mean $8 + j0$, $0 + j8$, $0 - j8$, or any value on a half circle of points in the *RX* plane that are 8 units away from the origin (Fig. 3-17). When an impedance *Z* is represented by the length (or absolute value) of its vector only, without specifying the reactance or resistance, *Z* is sometimes called an *absolute-value impedance*.

If an author (or an advertisement) does not tell us which complex impedance is meant when a single-number ohmic figure is quoted, the number usually refers to a *nonreactive impedance*. That means the imaginary, or reactive, factor is zero. This is also called a *resistive impedance* or a *pure resistance*.

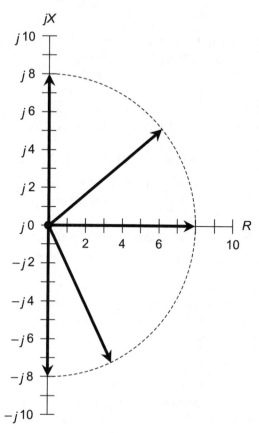

Fig. 3-17. Several different vectors, each theoretically representing $Z = 8\,\Omega$.

IMPEDANCES IN SERIES

Given two impedances $Z_1 = R_1 + jX_1$ and $Z_2 = R_2 + jX_2$ connected in series, the net complex impedance Z is their vector sum, given by:

$$Z = (R_1 + R_2) + j(X_1 + X_2)$$

For complex impedances in parallel, the process of finding the net impedance is more complicated, and is discussed later in this chapter.

PROBLEM 3-5

Suppose an inductor has a reactance $X_L = 45\,\Omega$ at a certain frequency, and a capacitor has a reactance $X_C = -55\,\Omega$ at that same frequency. If the two are connected in series, what is the net impedance, Z?

SOLUTION 3-5

The impedance of the inductor is $Z_L = 0 + j45$, and the impedance of the capacitor is $Z_C = 0 - j55$. When they are connected in series, the net impedance Z is:

$$Z = Z_L + Z_C$$
$$= (0 + j45) + (0 - j55)$$
$$= 0 + 0 + j(45 - 55)$$
$$= 0 + j(-10)$$
$$= 0 - j10$$

This is the equivalent of a pure capacitance whose reactance is $X_C = -10 \, \Omega$.

PROBLEM 3-6

Suppose the inductor in the above circuit changes so its reactance becomes $X_L = 55 \, \Omega$, while the reactance of the capacitor remains the same at $-55 \, \Omega$. Also suppose that the inductor has an internal DC resistance of $11 \, \Omega$. What is the net impedance, Z, when these two components are connected in series?

SOLUTION 3-6

In this case, the impedance of the inductor is $Z_L = 11 + j55$, and the impedance of the capacitor is $Z_C = 0 - j55$. Therefore, the net series impedance Z is found as follows:

$$Z = Z_L + Z_C$$
$$= (11 + j55) + (0 - j55)$$
$$= 11 + 0 + j(55 - 55)$$
$$= 11 + j0$$

This is a pure resistance of $R = 11 \, \Omega$. The capacitive and inductive reactances cancel each other out. This condition of equal and opposite reactances is called *resonance*. A circuit that contains both inductance and capacitance, and in which $X_L = -X_C$, is known as a *resonant circuit*.

RESONANT FREQUENCY

When a series circuit contains inductance and capacitance, resonance occurs at a specific frequency. Suppose the inductance (in henrys) is L, and the

capacitance (in farads) is C. Then the resonant frequency f_0 of the combination (in hertz) can be found according to this formula:

$$f_0 = 1 \Big/ \Big[2\pi (LC)^{1/2} \Big]$$

This formula also holds for f_0 in megahertz, L in microhenrys, and C in microfarads.

PROBLEM 3-7
Suppose an inductor whose value is 100 µH is connected in series with a capacitor whose value is 0.01 µF. What is the resonant frequency to the nearest hundredth of a megahertz?

SOLUTION 3-7
In this situation, $L = 100$ and $C = 0.01$, and we can use the above formula to calculate f_0:

$$f_0 = 1 \Big/ \Big[2\pi (100 \times 0.01)^{1/2} \Big]$$
$$= 1/(2\pi)$$
$$= 1/6.2832$$
$$= 0.16$$

This frequency, 0.16 MHz, can also be expressed as 160 kHz.

Characteristic Impedance

There is an interesting property of radio-frequency (RF) *transmission lines* that is often, somewhat imprecisely, called "impedance." This property is known as the *characteristic impedance* or *surge impedance*. It is symbolized Z_0 and can always be expressed as a positive real number. It's a scalar quantity, not a vector quantity. Within reasonable limits, the Z_0 of an RF transmission line is independent of the frequency, and is also independent of the length of the line. The Z_0 of a transmission line depends mainly on the cross-sectional dimensions and material composition.

TRANSMISSION LINES

An RF transmission line is used when it is necessary to get a signal from one place to another – usually from a wireless transmitter to an antenna or from an antenna to a wireless receiver – in an efficient manner. The line is supposed to carry the signal, which exists in the form of an EM field, without radiating any of it, and without picking up any external EM energy either.

The two most common types of RF transmission line are *parallel-wire line* and *coaxial cable*. Examples include the "twin lead" or "ribbon cable" that goes from an old-fashioned television (TV) receiving antenna to the TV set, and the cable used in community TV networks and Internet connections. The former type of line is called *balanced line* because its conductors carry equal and opposite RF currents. The latter type is called *unbalanced line* because its conductors carry different RF currents.

The Z_o of a parallel-wire transmission line depends on the diameter of the wires, on the spacing between the wires, and on the nature of the insulating, or *dielectric*, material separating the wires. The Z_o increases as the wire diameter gets smaller, and decreases as the wire diameter gets larger, all other things being equal. Typical TV "twin lead" has $Z_o = 300\,\Omega$.

In a coaxial line, as the center conductor is made thicker while the tubular outer conductor or *shield* stays the same size, the Z_o decreases. If the center conductor stays the same size and the shield increases in diameter, the Z_o increases. In general, the Z_o increases as the spacing between the center conductor and the shield increases. Solid dielectric materials such as polyethylene reduce the Z_o of a transmission line, compared with the Z_o when the same line has an air dielectric. Coaxial cables usually have Z_o values of $50\,\Omega$ or $75\,\Omega$.

STANDING WAVES

In antenna systems, zones of high and low RF current or voltage are common. These zones appear as *standing waves* that represent current or voltage maxima and minima. The waves are called "standing" because they remain in fixed positions unless some characteristic of the antenna changes. The subject of standing waves has given rise to a profusion of articles and books, as well as a few myths and misconceptions, among engineers and lay people.

Figure 3-18 shows three examples of standing waves on an antenna element. The pattern of standing waves on the radiating element of an antenna 1/2 wavelength long, fed at the center, is shown at Fig. 3-18A. The pattern of standing waves on a radiating element 1 wavelength long, fed at the center, is shown at B. The pattern of standing waves on a radiating element 1 wavelength long, fed at one end, is shown at C. These waves are caused by currents (solid lines) and voltages (dashed lines) reinforcing each other at specific points called *loops*, and canceling each other out at specific

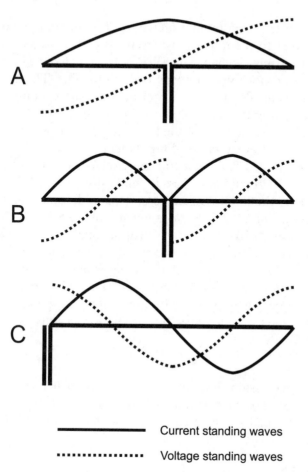

A

B

C

———————— Current standing waves

················ Voltage standing waves

Fig. 3-18.   Standing waves on a center-fed half-wave antenna (A), on a center-fed full-wave antenna (B), and on an end-fed full-wave antenna (C).

points called *nodes*. Standing waves on an antenna are normal and are no cause for concern. Standing waves on transmission lines, can, however, cause problems in some cases.

On an RF transmission line terminated in an impedance that is a pure resistance, and that is exactly the same as the Z_o of the line, there are no standing waves; the RF voltage E and the RF current I are uniform all along the line, and exist in the ratio $Z_o = E/I$. In this case there are no standing waves on the line. If the terminating impedance is not a pure resistance, or if it is a pure resistance but it differs from the Z_o of the line, a nonuniform distribution of current and voltage exists. The greater the *impedance mismatch*, the greater is the nonuniformity, and the more pronounced are the standing waves.

The ratio of the maximum voltage to the minimum voltage, or the maximum current to the minimum current, is called the *standing-wave ratio (SWR)* on a transmission line. The SWR is 1:1 when the current and voltage are in the same proportions everywhere along the line. The SWR can be 1:1 only when a transmission line is terminated in a non-reactive (purely resistive) load having the same ohmic value as the Z_o of the line. When this ideal scenario occurs, the ratio of the RF voltage to the RF current is equal to the Z_o of the line at every point along its length.

An extremely large SWR on a transmission line can cause significant loss of signal power. In any line, the loss is smallest when the SWR is 1:1. If the SWR is not 1:1, the line loss increases. This additional loss is called *SWR loss*, *impedance-mismatch loss*, or *feed-line mismatch loss*. Loss is measured in units called *decibels* (dB), which are proportional to the logarithm of the actual power loss. A loss of 1 dB represents the smallest decrease a listener can detect in a received signal, if that listener is expecting the change. A loss of 3 dB represents a loss of half the power. (Decibels are discussed in more detail in the chapter on amplifiers.) Figure 3-19 is a graph that approximately shows the feed-line mismatch loss that occurs for various values of matched-line loss and SWR.

PROBLEM 3-8
Suppose a transmission line has 3 dB of loss when perfectly matched (SWR = 1:1). Suppose the situation changes, so the SWR becomes 7:1. How much additional loss is introduced by this mismatch? What is the overall loss in the line when SWR = 7:1?

SOLUTION 3-8
Refer to Fig. 3-19. Locate the point for 3 dB on the horizontal axis. Proceed upward until the curve for the 7:1 SWR is reached. Then proceed horizontally to the left until the vertical axis is encountered. Read the SWR loss from the vertical scale. In this case it is approximately 2 dB. This means that the overall loss in the line is 3 dB + 2 dB = 5 dB.

Admittance

Admittance is a measure of the ease with which a medium carries AC. It is the AC counterpart of DC conductance, but, like AC impedance, it is a complex quantity. Admittance has a real-number part that represents conductance, and an imaginary-number part that represents *susceptance*.

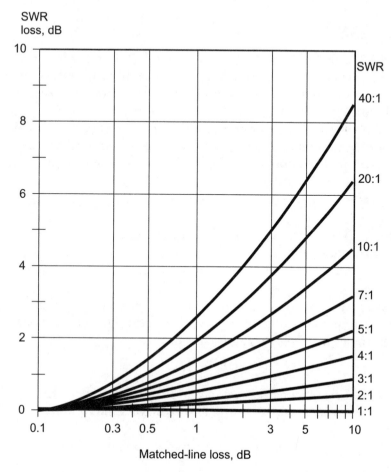

Fig. 3-19.   SWR losses produced by impedance mismatches on an RF transmission line.

AC CONDUCTANCE

In an AC circuit, electrical conductance works the same way as it does in a DC circuit. Conductance is symbolized by the capital letter G. The relationship between conductance and resistance is simple:

$$G = 1/R$$

The unit of conductance is the *siemens* (called the *mho* in archaic documents), abbreviated by the non-italicized, uppercase letter S. As conductance increases, resistance goes down and current (I) goes up. Conversely, as G decreases, R goes up and I goes down.

SUSCEPTANCE

Susceptance is symbolized by the capital letter B. It is the reciprocal of AC reactance. Susceptance, like reactance, can be either capacitive or inductive. Capacitive susceptance is symbolized B_C, and inductive susceptance is symbolized B_L.

The expression of susceptance requires the j operator, just as does the expression of reactance. The j operator behaves strangely when it is the denominator in a fraction or a quotient. The reciprocal of j is the same as its negative! That is:

$$1/j = -j$$

Therefore, the reciprocal of an imaginary number jX, where X is the real-number multiple of j, is:

$$1/(jX) = -j(1/X)$$

When you want to find susceptance in terms of reactance, first change the sign of j, and then take the reciprocal of the real-number component.

The formula for *capacitive susceptance* is:

$$B_C = 2\pi fC$$

where B_C is the capacitive susceptance in siemens, f is the frequency in hertz, and C is the capacitance in farads. This resembles the formula for inductive reactance.

The formula for *inductive susceptance* is similar to that for capacitive reactance:

$$B_L = -1/(2\pi fL)$$

where B_L is the inductive susceptance in siemens, f is the frequency in hertz, and L is the inductance in henrys. In the case of an inductive susceptance, we have "negative siemens," something that you never hear about with DC conductance.

COMPLEX ADMITTANCE

Admittance is the complex composite of conductance and susceptance:

$$Y = G + jB$$

The j factor is sometimes negative, so sometimes we write $Y = G - jB$. If we have a component whose reactance is $Z = R + jX$, its admittance is

$Y = 1/R - j(1/X)$. Given a component whose reactance is $Z = R - jX$, its admittance is $Y = 1/R + j(1/X)$.

Admittance, rather than impedance, is preferable when making calculations involving parallel AC circuits. Resistance and reactance combine in messy fashion in parallel circuits, but conductance (G) and susceptance (B) simply add together, yielding admittance (Y). The situation is similar to the behavior of resistances in parallel when working with DC circuits.

THE *GB* PLANE

Admittance can be depicted on a plane that looks like the complex impedance (RX) plane. Conductance is plotted along the horizontal, or G, axis, and susceptance is plotted along the B axis (Fig. 3-20).

The center, or *origin*, of the *GB plane* represents the point at which there is no conduction for DC or AC. In the RX plane, the origin represents a perfect

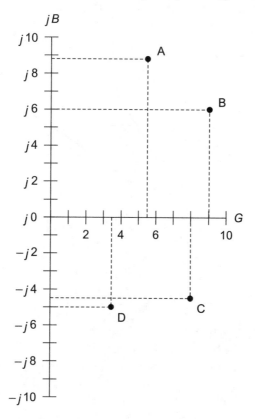

Fig. 3-20. Some points in the *GB* plane, and their components on the G and B axes. At A, $5.5 + j8.8$. At B, $9 + j6$. At C, $8 - j4.5$. At D, $3.4 - j5$.

short circuit; in the *GB* plane the origin corresponds to a perfect open circuit. An open circuit in the *RX* plane is represented by points infinitely far from the origin in all directions. In the *GB* plane, these points represent a short circuit. The *GB* plane is, in a mathematical sense, blown inside-out compared with the *RX* plane.

VECTOR REPRESENTATION OF ADMITTANCE

Complex admittances can be shown as vectors, just as can complex impedances. In Fig. 3-21, the points from Fig. 3-20 are rendered as vectors. Longer vectors indicate, in general, greater flow of current, and shorter ones indicate less current.

Vectors pointing generally "northeast," or upwards and to the right, represent conductances and capacitances in parallel. Vectors pointing in a more or less "southeasterly" direction, or downwards and to the right, represent conductances and inductances in parallel.

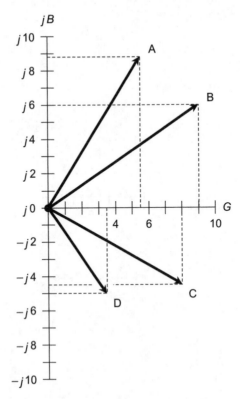

Fig. 3-21.   Some vectors in the *GB* plane. At A, $5.5+j8.8$. At B, $9+j6$. At C, $8-j4.5$. At D, $3.4-j5$.

ADMITTANCES IN PARALLEL

Given two admittances $Y_1 = G_1 + jB_1$ and $Y_2 = G_2 + jB_2$ connected in parallel, the net admittance Y is their vector sum, given by:

$$Y = (G_1 + G_2) + j(B_1 + B_2)$$

When it is necessary to find the net impedance of two components in parallel, first convert each resistance to conductance, and each reactance to suscep-tance. Combine these to get the admittances. Use the above formula to find the net admittance. Then convert the net conductance back to resistance, and the net susceptance back to reactance. The resulting composite is the net impedance of the components in parallel.

PROBLEM 3-9

What is the complex admittance Y_1 of a component whose impedance is $Z_1 = 10 + j100$? What is the complex admittance Y_2 of a component whose impedance is $Z_2 = 10 - j100$?

SOLUTION 3-9

Both admittances are found by taking the reciprocals of both components of the impedances. Mathematically:

$$Y_1 = 1/10 - j(1/100)$$
$$= 0.1 - j0.01$$

$$Y_2 = 1/10 + j(1/100)$$
$$= 0.1 + j0.01$$

PROBLEM 3-10

Suppose the above two components are connected in parallel. What is the resulting admittance Y? What is the resulting impedance Z?

SOLUTION 3-10

The composite admittance, Y, is found by adding the two admittances as vectors:

$$Y = Y_1 + Y_2$$
$$= 0.1 - j0.01 + 0.1 + j0.01$$
$$= (0.1 + 0.1) + j(-0.01 + 0.01)$$
$$= 0.2 + j0$$

This represents a conductance of $0.2\,\text{S}$ and a susceptance of $0\,\text{S}$.

The composite impedance, Z, is found by taking the reciprocals of the conductance and susceptance. Thus:

$$Z = 1/0.2 + 1/(j0)$$
$$= 5 - j(1/0)$$

The multiple of j is mathematically undefined here. In practice, a zero susceptance behaves like an "infinite reactance" or open circuit. With the effective resistance of $5\,\Omega$ in parallel with the equivalent of an open circuit, this arrangement behaves like a $5\,\Omega$ resistor under the conditions specified, and is called *parallel-resonant* because it contains equal and opposite reactances in parallel. Parallel resonance always occurs at a defined frequency, which can be found using the same formula as the one that applies to series resonance, given earlier in this chapter.

Quiz

Refer to the text in this chapter if necessary. Answers are in the back of the book.

1. What is the approximate reactance of the capacitor shown in Fig. 3-22? Assume the component values are exact.
    (a) $200\,\Omega$
    (b) $-0.005\,\Omega$
    (c) $-200\,\Omega$
    (d) More information is needed to answer this question.

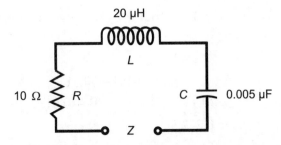

Fig. 3-22.   Illustration for Quiz Questions 1 through 3.

2. At a frequency of 10,000 kHz, what is the approximate complex
    impedance of the circuit shown in Fig. 3-22? Assume the component
    values are exact.
    (a) $10 + j1253$
    (b) $10 - j3.183$
    (c) $10 + j19.995$
    (d) More information is needed to answer this question.

3. What is the approximate resonant frequency of the circuit shown in
    Fig. 3-22? Assume the component values are exact.
    (a) 503 kHz
    (b) 1.987 MHz
    (c) 1987 MHz
    (d) More information is needed to answer this question.

4. In order to add two complex numbers,
    (a) the real parts and the imaginary parts must be multiplied.
    (b) the real parts and the imaginary parts must be added separately.
    (c) the real and imaginary parts must be canceled out.
    (d) the reciprocals of both must be taken first.

5. Suppose there are two RF transmission lines, called line A and line B.
    Both are of the parallel-wire type. In both instances, the spacing
    between conductors is 25 millimeters (mm), and the dielectric material
    separating the conductors is a flat, continuous ribbon of solid
    polyethylene. In line A, the wires are 1.00 mm in diameter, while in
    line B, the wires are 2.00 mm in diameter. Which of the following
    statements is true?
    (a) The Z_o of line A is greater than the Z_o of line B.
    (b) The Z_o of line B is greater than the Z_o of line A.
    (c) The Z_o of line A is the same as the Z_o of line B.
    (d) We need more information to say how the Z_o values compare.

6. In a resonant circuit, the susceptances
    (a) are equal and similar (both inductive or both capacitive).
    (b) are equal and opposite (one inductive and the other capacitive).
    (c) are both zero.
    (d) are both undefined.

7. The characteristic impedance of a coaxial transmission line depends
    on all of the following except
    (a) the diameter of the wire inner conductor
    (b) the diameter of the tubular shield or outer conductor

(c) the line length

(d) the dielectric material separating the center conductor from the shield

8. When a complex number $R + jX$ is represented as a point in the RX plane, the absolute value of that complex number is equal to
   (a) the R coordinate.
   (b) the X coordinate.
   (c) the negative of the X coordinate.
   (d) None of the above.

9. What is the susceptance of a 1000 µH inductor at a frequency of 100 Hz? Express the answer to the nearest hundredth of a siemens.
   (a) -0.001 S
   (b) -0.63 S
   (c) -1.59 S
   (d) -1000 S

10. Suppose a component has a complex impedance $Z = R + jX$, where X is inductive and $X = 2R$. What is the phase angle to the nearest degree?
   (a) It cannot be calculated without more information.
   (b) $0°$
   (c) $37°$
   (d) $63°$

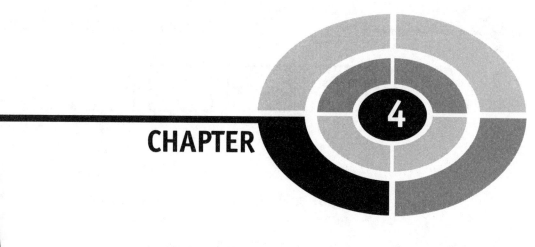

CHAPTER 4

Power Supplies

Most electronic equipment requires DC. In the United States, the electricity from the utility company is 60 Hz AC at 117 V or 234 V rms. A power supply converts the utility AC to DC having a predetermined voltage.

Power supplies can be dangerous. *If you have any doubt about your ability to safely work with an electrical power supply, then leave it to a professional.* Figure 4-1 is a block diagram of a DC power supply. Let's learn about each of the major sections, or stages.

Transformers

Power-supply transformers are available in two types: the *step-down transformer* that decreases the AC voltage, and the *step-up transformer* that increases the AC voltage. The AC output frequency from a transformer is always the same as the AC input frequency.

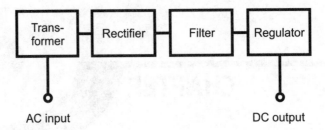

Fig. 4-1. Block diagram of a basic DC power supply.

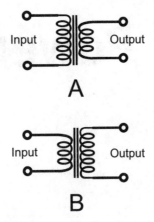

Fig. 4-2. At A, the schematic symbol for a step-down power transformer. At B, the schematic symbol for a step-up power transformer.

STEP-DOWN TRANSFORMER

Most electronic devices, such as radio receivers, computers, and camcorders, need only a few volts. The power supplies for such equipment use step-down power transformers (Fig. 4-2A). The physical size of the transformer depends on the current it is required to deliver. The *input-to-output voltage ratio* is directly proportional to the *primary-to-secondary turns* ratio. That is:

$$V_{in}/V_{out} = N_{pri}/N_{sec},$$

where V_{in} is the rms input voltage, V_{out} is the rms output voltage, N_{pri} is the number of turns in the primary winding, and N_{sec} is the number of turns in the secondary winding.

The transformer in a low-current system such as a radio receiver can be physically small. But high-current devices such as large audio amplifiers or amateur-radio transmitters need bulkier, heavier transformers. This is because in a high-current system, the transformer secondary winding must be of heavy-gauge wire, and the *core*, or material on which the wire is wound, must be massive.

STEP-UP TRANSFORMER

Some circuits require high voltage (more than 117 V DC). The cathode-ray tube (CRT) in an old-fashioned television receiver needs several hundred volts. Some amateur-radio power amplifiers use vacuum tubes working at several kilovolts DC. The transformers in these appliances are step-up types.

The voltage-transformation formula for a step-up transformer is the same as the formula given above for step-down transformers. The input-to-output voltage ratio is, again, the same as the primary-to-secondary turns ratio.

If a step-up transformer is required to supply a small current, it need not be large physically. But some step-up transformers are used with vacuum-tube equipment that consumes considerable power, such as broadcast transmitting amplifiers. For this reason, step-up transformers are generally larger than the step-down units in low-voltage systems such as computers.

TRANSFORMER RATINGS

Transformers are rated according to output voltage and current. For a given unit, the *volt-ampere (VA) capacity* is often specified. This is the product of the voltage and current. A transformer with a 12 V output, capable of delivering 10 A, would have 12 V × 10 A = 120 VA of capacity.

The nature of power-supply filtering, discussed later in this chapter, makes it necessary for the power-transformer VA rating to be greater than the actual power, in watts, consumed by the load.

A rugged power transformer, capable of providing the necessary current and/or voltage on a continuous basis, is crucial in any power supply. The transformer is usually the most expensive component to replace in the event of a power-supply failure.

PROBLEM 4-1
Suppose you need a transformer that will provide 12 V rms AC output when supplied with 120 V rms AC input. Do you need a step-down type or a step-up type?

SOLUTION 4-1
Because the output voltage is lower than the input voltage, you need a step-down transformer.

PROBLEM 4-2

What should the primary-to-secondary turns ratio be in the situation described by the previous problem?

SOLUTION 4-2

The primary-to-secondary turns ratio (N_{pri}/N_{sec}) must be the same as the input-to-output voltage ratio (V_{in}/V_{out}). We are told that $V_{in} = 120$ and $V_{out} = 12$. Therefore, $V_{in}/V_{out} = 120/12 = 10$, meaning that $N_{pri}/N_{sec} = 10$. Sometimes this is written $10:1$ (read "10 to 1") to emphasize the fact that it is a ratio.

Rectifiers

A *rectifier* circuit converts AC to *pulsating DC*. This is usually accomplished by means of one or more heavy-duty *semiconductor diodes* following a power transformer.

HALF-WAVE CIRCUIT

The simplest type of rectifier circuit, known as the *half-wave rectifier* (Fig. 4-3A), uses one diode (or a series or parallel combination of diodes if high current or voltage is required) to "chop off" half of the AC cycle. The rms output voltage in this type of circuit is approximately 45% of the applied rms AC voltage (Fig. 4-4A). But the peak voltage in the reverse direction, called the *peak inverse voltage* (PIV) or *peak reverse voltage* (PRV) across the diode, can be as much as 2.8 times the applied rms AC voltage. Most engineers like to use diodes whose PIV ratings are at least 1.5 times the maximum expected PIV. Thus, with a half-wave supply, the diodes should be rated for at least 2.8×1.5, or 4.2, times the rms AC voltage that appears across the secondary winding of the power transformer.

Half-wave rectification has shortcomings. First, the output is difficult to filter. Second, the output voltage can diminish considerably when the supply is required to deliver high current. Third, half-wave rectification puts a strain on the power transformer and the diodes because it "pumps" them. It lets them loaf during half the AC cycle, but works them hard during the other half.

Half-wave rectification usually works all right in a power supply that is not required to deliver much current, or when the voltage can vary without

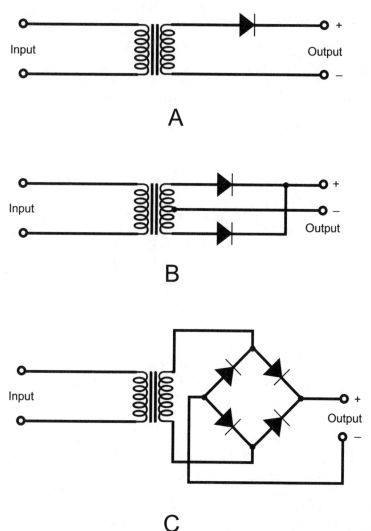

Fig. 4-3. At A, a half-wave rectifier circuit. At B, a full-wave center-tap rectifier circuit. At C, a full-wave bridge rectifier circuit.

affecting the behavior of the equipment connected to it. The main advantage of a half-wave circuit is that it costs less than more sophisticated circuits.

FULL-WAVE CENTER-TAP CIRCUIT

A better scheme for changing AC to DC takes advantage of both halves of the AC cycle. A *full-wave center-tap rectifier* has a transformer with a tapped

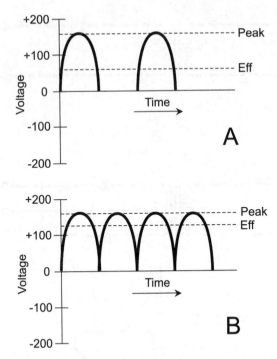

Fig. 4-4. At A, the output of a half-wave rectifier. At B, the output of a full-wave rectifier. The effective (eff) or rms voltages are always less than the peak voltages.

secondary, as shown in Fig. 4-3B. The center tap is connected to *electrical ground*. This produces voltages and currents at the ends of the winding that are in phase opposition with respect to each other. These two AC waves can be individually half-wave rectified, cutting off one half of the cycle and then the other, over and over.

In this circuit, the average DC output voltage is about 90% of the rms AC input voltage (Fig. 4-4B). The PIV across the diodes can nevertheless be as much as 2.8 times the applied rms AC voltage. Therefore, the diodes should have a PIV rating of at least 4.2 times the applied rms AC voltage to ensure that they won't break down.

The output of a full-wave rectifier is easier to filter than that of a half-wave rectifier. The full-wave rectifier is easier on the transformer and diodes than a half-wave circuit. If a load is applied to the output of the full-wave circuit, the voltage drops less than is the case with a half-wave supply. In all these respects, the full-wave circuit is superior to the half-wave circuit. But because the transformer is more sophisticated, the full-wave circuit costs more than a half-wave circuit that delivers the same output voltage.

FULL-WAVE BRIDGE CIRCUIT

Another way to get full-wave rectification is the *full-wave bridge rectifier*, often called simply a *bridge*. It is diagrammed in Fig. 4-3C. The output waveform is similar to that of the full-wave center-tap circuit (Fig. 4-4B).

The average DC output voltage in the bridge circuit is 90% of the applied rms AC voltage, as is the case with center-tap rectification. The PIV across the diodes is 1.4 times the applied rms AC voltage. Therefore, each diode needs to have a PIV rating of at least 1.4×1.5, or 2.1, times the rms AC voltage that appears at the transformer secondary.

The bridge circuit does not need a center-tapped transformer secondary. It uses the entire secondary winding on both halves of the wave cycle. The bridge circuit therefore makes more efficient use of the transformer. The bridge is also easier on the diodes than half-wave or full-wave center-tap circuits.

The main disadvantage of the bridge is that it needs four diodes rather than two. This does not always amount to much in terms of cost, but it can be important when a power supply must deliver high current. Then, the extra diodes – two for each half of the cycle, rather than one – dissipate more overall heat energy.

VOLTAGE-DOUBLER CIRCUIT

Diodes and capacitors can be connected in certain ways to cause a power supply to deliver an output that is a multiple of the peak AC input voltage. Theoretically, large whole-number multiples are possible. But it is rare to see power supplies that make use of multiplication factors larger than 2.

In practice, *voltage-multiplier power supplies* work well only when the load draws low current. Otherwise, the *voltage regulation* is poor; the voltage drops a lot when the current demand is significant. In high-current, high-voltage applications, the best way to build a power supply is to use a step-up transformer, not a voltage-multiplication scheme.

A *voltage-doubler circuit* is shown in Fig. 4-5. This circuit works on the entire AC cycle, and is called a *full-wave voltage doubler*. Its DC output voltage, when the current drain is low, is about twice the peak AC input voltage, or about 2.8 times the applied rms AC voltage. This circuit subjects the diodes to a PIV of 2.8 times the applied rms AC voltage. Therefore, they should be rated for PIV of at least 4.2 times the rms AC voltage that appears across the transformer secondary.

Proper operation of this type of circuit depends on the ability of the capacitors to hold a charge under maximum load. This means that the

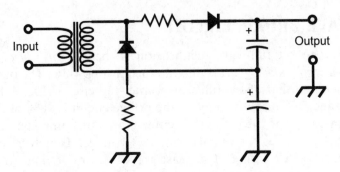

Fig. 4-5. A full-wave voltage-doubler power supply.

capacitors must have large values, as well as be capable of handling high voltages. The capacitors serve two purposes: to boost the voltage and to filter the output. The resistors, which have low ohmic values, protect the diode against *surge currents* that occur when the power supply is first switched on.

PROBLEM 4-3

Suppose a power transformer with a 1:4 primary-to-secondary turns ratio is connected to the 120 V AC utility mains. A half-wave rectifier circuit is used to obtain pulsating DC output. What is the minimum PIV rating the diodes should have in order to ensure that they won't break down?

SOLUTION 4-3

The utility mains supply 120 V rms AC, and the transformer steps up this voltage by a factor equal to the primary-to-secondary turns ratio. Therefore, the output of the transformer is 4 × 120 V rms AC, or 480 V rms AC. In a half wave circuit, the PIV rating of the diodes should be at least 4.2 times this. That means they must be rated for at least 4.2 × 480 PIV, or 2016 PIV.

PROBLEM 4-4

Suppose a full-wave bridge rectifier circuit is used in the above scenario, rather than a half-wave rectifier circuit. What must be the minimum PIV rating of the diodes in this case?

SOLUTION 4-4

The transformer secondary still delivers 480 V rms AC. In a full-wave circuit, the PIV rating of the diodes should be at least 2.1 times the rms AC voltage at the transformer secondary. This is half the PIV in the half-wave situation: 2.1 × 480 PIV, or 1008 PIV.

Filtering and Regulation

Most electronic equipment requires something better than the pulsating DC that comes straight from a rectifier circuit. The pulsation, called *ripple*, in the rectifier output is minimized or eliminated by a *filter*.

CAPACITORS ALONE

The simplest power-supply filter consists of one or more large-value capacitors, connected in parallel with the rectifier output (Fig. 4-6). A good component for this purpose is the *electrolytic capacitor*. It is *polarized*, meaning that it must be connected in the correct direction. Typical values range in the hundreds, or even in the thousands, of microfarads. The more current that a power supply must deliver, the more capacitance is needed for effective filtering. This is because larger capacitances hold charge for a longer time with a given current load than do smaller capacitances.

Filter capacitors work by "trying" to maintain the DC voltage at its peak level (Fig. 4-7). This is easier to do with the output of a full-wave rectifier (A) than with the output of a half-wave rectifier (B). With a full-wave rectifier receiving a 60 Hz AC electrical input, the ripple frequency is 120 Hz; with a half-wave rectifier it is 60 Hz. The filter capacitors are recharged twice as often with a full-wave rectifier, as compared with a half-wave rectifier. The result is that the ripple is less severe, for a given filter capacitance, when a full-wave circuit is used.

CAPACITORS AND CHOKES

Another way to smooth out the DC from a rectifier is to place a large-value inductor in series with the output, and a large-value capacitance in parallel. The inductor, called a *filter choke*, has a value of up to several henrys.

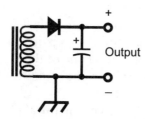

Fig. 4-6.   A simple capacitor power-supply filter.

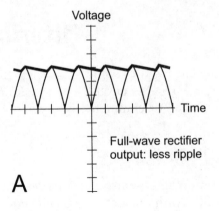

A

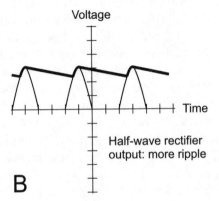

B

Fig. 4-7. Filtering of ripple from a full-wave rectifier (A) and from a half-wave rectifier (B).

Sometimes the capacitor is placed on the rectifier side of the choke. This circuit is a *capacitor-input filter* (Fig. 4-8A). If the filter choke is placed on the rectifier side of the capacitor, the circuit is a *choke-input filter* (Fig. 4-8B). Capacitor-input filtering can be used when a power supply is not required to deliver much current. The output voltage is higher with a capacitor-input circuit than with a choke-input circuit. If the supply needs to deliver large or variable amounts of current, a choke-input filter is a better choice, because the output voltage is more stable.

If the output of a DC power supply must have an absolute minimum of ripple, two or three capacitor/choke pairs can be connected in *cascade* (Fig. 4-9). Each pair constitutes a *section* of the filter. Multi-section filters can consist of either capacitor-input or choke-input sections, but the two types are never mixed.

In the example of Fig. 4-9, both capacitor/choke pairs are called *L sections* because of their arrangement in the schematic diagram. If the second

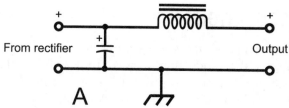

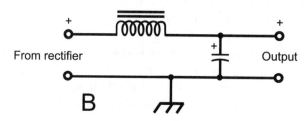

Fig. 4-8.   At A, capacitor-input filter. At B, choke-input filter.

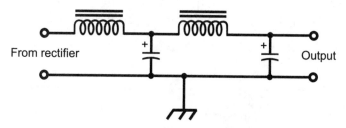

Fig. 4-9.   Two choke-input filter sections in cascade.

capacitor is omitted, the filter becomes a *T section*. If the second capacitor is moved to the input and the second choke is omitted, the filter becomes a *pi section*. These sections are named because their schematic diagrams look something like the uppercase English L, the uppercase English T, and the uppercase Greek Π respectively.

PROBLEM 4-5

Suppose the standard AC line frequency is higher than 60 Hz. Does this make the output of a rectifier circuit easier to filter, or more difficult?

SOLUTION 4-5

In theory, it makes the output of a rectifier circuit easier to filter, because the capacitors don't have to hold the charge for as long.

VOLTAGE REGULATION

If a special diode called a *Zener diode* is connected in parallel with the output of a power supply, the diode limits the output voltage. The diode must have an adequate power rating to prevent it from burning out. The limiting voltage depends on the particular Zener diode used. There are Zener diodes available that will fit any reasonable power-supply voltage. Figure 4-10 is a diagram of a full-wave, center-tapped DC power supply including a Zener diode for voltage regulation. Note the direction in which the Zener diode is connected. It's important that the diode polarity be correct, or it will burn out.

A Zener-diode voltage regulator is inefficient when the supply is used with equipment that draws high current. When a supply must deliver a lot of current, a *power transistor* is used along with the Zener diode to obtain regulation. A circuit diagram of such a scheme is shown in Fig. 4-11.

Voltage regulators are available in *integrated-circuit* (IC) form. Such an IC, sometimes along with some external components, is installed in the power-supply circuit at the output of the filter. In high-voltage power supplies, *electron-tube voltage regulators* are sometimes used.

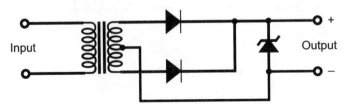

Fig. 4-10. A full-wave center-tap supply with a Zener-diode regulator.

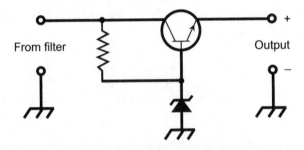

Fig. 4-11. A regulator circuit using a Zener diode and an NPN transistor.

Protection of Equipment

The output of a power supply should be free of sudden changes that can damage equipment or components, or interfere with their proper performance. It is also important that voltages do not appear on the external surfaces of a power supply, or on the external surfaces of any equipment connected to it.

GROUNDING

The best electrical ground for a power supply is the "third wire" ground provided in up-to-date AC utility circuits. The "third hole" (the bottom hole in an AC outlet, shaped like an uppercase English letter D turned on its side) should be connected to a wire that ultimately terminates in a *ground rod* driven into the earth at the point where the electrical wiring enters the building.

In older buildings, *two-wire AC systems* are common. These can be recognized by the presence of only two slots in the utility outlets. Some of these systems employ reasonable grounding by means of a scheme called *polarization*, where one slot is longer than the other, the longer slot being connected to electrical ground. But this is not as good as a *three-wire AC system*, in which the ground connection is independent of both the outlet slots.

Unfortunately, the presence of a three-wire or polarized outlet system does not always mean that an appliance connected to an outlet is well grounded. If the appliance design is faulty, or if the "third hole" was not grounded by the people who installed the electrical system, a power supply can deliver unwanted voltages to the external surfaces of appliances and electronic devices. This can present an electrocution hazard, and can also hinder the performance of electronic equipment.

Warning: All metal chassis and exposed metal surfaces of AC power supplies should be connected to the grounded wire of a three-wire electrical cord. The "third prong" of the plug should never be defeated or cut off. Some means should be found to ensure that the electrical system in the building has been properly installed, so you don't work under the illusion that your system has a good ground when it really doesn't. If you're in doubt about this, consult a professional electrician.

SURGE CURRENTS

At the instant a power supply is switched on, a surge of current occurs, even with nothing connected to the supply output. This is because the filter

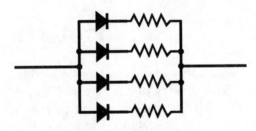

Fig. 4-12.   Diodes in parallel, with current-equalizing resistors in series with each diode.

capacitors need an initial charge, so they draw a large current for a short time. The surge current is far greater than the normal operating current. An extreme surge can destroy the rectifier diodes in a power supply. The phenomenon is worst in high-voltage supplies and voltage-multiplier circuits. Diode failure as a result of current surges can be prevented in at least three ways:

- Use diodes with a current rating of many times the normal operating level.
- Connect several diodes in parallel wherever a diode is called for in the circuit. *Current-equalizing resistors* are necessary (Fig. 4-12). The resistors should have a small ohmic value. The diodes should all be identical.
- Use an automatic switching circuit in the transformer primary. This applies a reduced AC voltage for 1 or 2 seconds, and then applies full input voltage.

TRANSIENTS

The AC on the utility line is a sine wave with a constant rms voltage near 117 V or 234 V. But there are often *voltage spikes*, known as *transients*, that can attain peak values of several thousand volts. Transients are caused by sudden changes in the load in a utility circuit. A heavy thundershower can produce transients throughout an entire town. Unless they are suppressed, transients can destroy the diodes in a power supply. Transients can also befuddle the operation of sensitive electronic equipment such as computers or microcomputer-controlled appliances.

The simplest way to get rid of common transients is to place a capacitor of about 0.01 µF, rated for 600 V or more, between each side of the transformer primary and electrical ground (Fig. 4-13). Commercially made *transient suppressors* are available. These devices, often mistakenly called "surge protectors," use more sophisticated methods to prevent sudden voltage

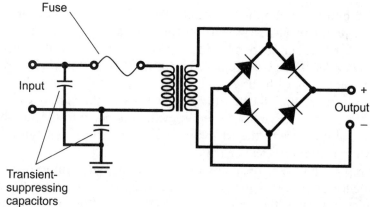

Fig. 4-13.   A power supply with transient-suppression capacitors and a fuse in the
transformer primary circuit.

spikes from reaching levels where they can cause problems. It is a good idea
to use transient suppressors with all sensitive electronic devices, including
computers, hi-fi stereo systems, and television sets. In the event of a
thundershower, the best way to protect such equipment is to physically
unplug it from the wall outlets until the storm has passed.

FUSES

A *fuse* is a piece of soft wire that melts, breaking a circuit if the current
exceeds a certain level. A fuse is placed in series with the transformer primary
(shown in Fig. 4-13 along with transient-suppressing capacitors). A short
circuit or overload will burn the fuse out. If a fuse blows out, it must be
replaced with another of the same rating.

Fuses are available in two types: the *quick-break fuse* and the *slow-blow
fuse.* A quick-break fuse is a straight length of wire or a metal strip. A slow-
blow fuse usually has a spring inside along with the wire or strip. It's best
to replace blown-out fuses with new ones of the same type. Quick-break
fuses in slow-blow situations can burn out needlessly, causing inconvenience.
Slow-blow fuses in quick-break environments might not provide adequate
protection to the equipment, letting excessive current flow for too long before
blowing out.

CIRCUIT BREAKERS

A *circuit breaker* performs the same function as a fuse, except that a breaker
can be reset by turning off the power supply, waiting a moment, and then

pressing a button or flipping a switch. Some breakers reset automatically when the equipment has been shut off for a certain length of time.

If a fuse or breaker keeps blowing out or tripping, or if it blows or trips immediately after it has been replaced or reset, then something is wrong with the power supply or with the equipment connected to it. Burned-out diodes, a bad transformer, and shorted filter capacitors in the supply can all cause this trouble. A short circuit in the equipment connected to the supply, or the connection of a device in the wrong direction (polarity), can cause repeated fuse blowing or circuit-breaker tripping.

Never replace a fuse or breaker with a larger-capacity unit to overcome the inconvenience of repeated fuse/breaker blowing/tripping. Find the cause of the trouble, and repair the equipment as needed. The "penny in the fuse box" scheme can endanger equipment and personnel, and it increases the risk of fire in the event of a short circuit.

PROBLEM 4-6
Will a transient suppressor work properly if it is designed for a three-wire electrical system, but the ground wire has been defeated, cut off, or does not lead to a good electrical ground?

SOLUTION 4-6
No. In order to properly function, a transient suppressor needs a substantial ground. The excessive, sudden voltages are shunted away from sensitive equipment only when a current path is provided to allow discharge to ground.

Electrochemical Power Sources

An *electrochemical cell*, often called simply a *cell*, is a source of DC power that can exist independently of utility systems. When two or more cells are connected in series, the result is a *battery*. Cells and batteries are used in portable electronic equipment, in communications satellites, and as sources of emergency power.

ELECTROCHEMICAL ENERGY

Figure 4-14 is a simplified drawing of a *lead-acid cell*. An electrode made of lead and another electrode made of lead dioxide, immersed in a sulfuric-acid

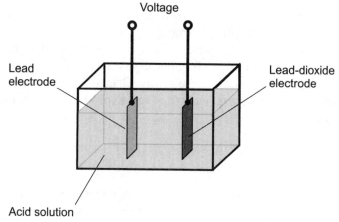

Fig. 4-14.   A lead-acid cell.

solution, acquire a potential difference. This voltage can drive a current through a load. The maximum available current depends on the volume and mass of the cell.

If this cell is connected to a load for a long time, the current will gradually decrease, and the electrodes will become coated. The nature of the acid will change. All the potential energy in the acid will be converted into DC electrical energy. However, the cell can be recharged by connecting it to an external source of voltage equal to approximately the normal cell voltage, positive-to-positive and negative-to-negative. The recharging process depends on the volume and mass of the cell, and can take up to several hours to complete.

PRIMARY AND SECONDARY CELLS

Some cells, once their chemical energy has all been changed to electricity and used up, must be discarded. These are *primary cells*. Other kinds of cells, such as the lead-acid unit described above, can get their chemical energy back again by recharging. Such a component is a *secondary cell*.

Primary cells contain a dry electrolyte paste along with metal electrodes. They go by names such as *dry cell*, *zinc-carbon cell*, or *alkaline cell*, and are commonly found in supermarkets and department stores. Some secondary cells can also be found in stores. *Lithium cells* are one common type. These cost more than ordinary dry cells, and a charging unit also costs a few dollars. But these rechargeable cells can be used hundreds of times, and can pay for themselves and the charger several times over.

An *automotive battery* is made from secondary cells connected in series. These cells recharge from the alternator or from an outside charging unit. This battery has cells like the one shown in Fig. 4-14. It is dangerous to short-circuit the terminals of such a battery, because the acid can boil out. (In fact, it is unwise to short-circuit any electrochemical cell or battery, because under certain circumstances they can explode or heat up to the point that they cause fires.)

STANDARD CELL

Most cells produce between 1.0 V and 1.8 V DC. Some types of cells generate predictable and precise voltages. These are known as *standard cells*. One example is the *Weston standard cell*. It produces 1.018 V when operated at room temperature (approximately 20°C or 68°F). This type of cell has an electrolyte solution of cadmium sulfate. The positive electrode is mercury sulfate, and the negative electrode consists of mercury and cadmium (Fig. 4-15).

When a Weston standard cell is properly constructed and used at room temperature, its voltage is always the same. This allows it to be employed as a DC voltage reference.

STORAGE CAPACITY

Common units of electrical energy are the *watt hour* (Wh) and the *kilowatt hour* (kWh), where 1 kWh = 1000 Wh. Any cell or battery has a certain amount of electrical energy that can be specified in watt hours or kilowatt

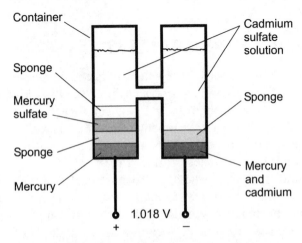

Fig. 4-15. A Weston standard cell.

hours. Often it is given in terms of the mathematical integral of deliverable current with respect to time, in units of *ampere hours* (Ah). The *storage capacity* in watt hours is the ampere-hour capacity multiplied by the battery voltage.

A battery with a rating of 20 Ah can provide 20 A for 1 h, or 1 A for 20 h, or 100 mA (100 milliamperes) for 200 h. The extreme situations are *shelf life* and *maximum deliverable current*. Shelf life is the length of time the battery will remain usable if it is never connected to a load; this can be months or years. The maximum deliverable current is the highest current a battery can drive through a load without the voltage dropping significantly because of the battery's internal resistance.

Small cells have storage capacities on the order of a few milliampere hours (mAh) up to 100 or 200 mAh. Medium-sized cells supply 500 mAh to 1000 mAh (1 Ah). Large automotive lead-acid batteries can provide upwards of 100 Ah.

PROBLEM 4-7
Suppose a battery can supply 10 Ah at 12 V DC. How many watt hours (Wh) does this represent?

SOLUTION 4-7
Recall the formula for DC power in terms of voltage and current:

$$P = EI$$

where P is the power in watts, E is the voltage in volts, and I is the current in amperes. Multiplying this through by time t, in hours, gives us this formula:

$$Pt = EIt$$

In this equation, the left-hand side represents watt hours, while the right-hand side represents the product of volts and ampere hours. In this particular situation, $E = 12$ and $It = 10$. Therefore:

$$
\begin{aligned}
Pt &= EIt \\
&= 12 \times 10 \\
&= 120 \text{ Wh}
\end{aligned}
$$

DISCHARGE CURVES

When an *ideal cell* or *ideal battery* (that is, a theoretically perfect unit that operates exactly according to a certain set of specifications) is placed

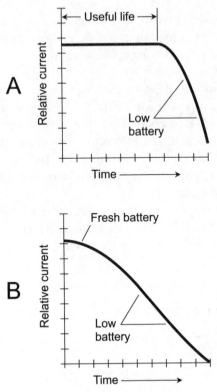

Fig. 4-16. At A, a flat discharge curve. This is the ideal for battery performance. At B, a declining discharge curve.

in service, it is expected to deliver a constant current for awhile. Then the current is expected to decrease. Some types of cells and batteries approach ideal behavior, exhibiting a *flat discharge curve* (Fig. 4-16A). Less optimal cells and batteries deliver current that decreases gradually from the beginning of use; this is known as a *declining discharge curve* (Fig. 4-16B).

When the current that a cell or battery can provide has decreased to significantly less than its initial value, the cell or battery is said to be "weak" or "low." At this time, it should be replaced. If it is allowed to run down until the current drops to nearly zero, the cell or battery is said to be "dead" (although, in the case of a rechargeable unit, a better term might be "sleeping").

COMMON CELLS AND BATTERIES

The cells sold in stores, and used in convenience items like flashlights and transistor radios, are usually of the zinc-carbon or alkaline variety. These provide

1.5 V and are available in sizes AAA (very small), AA (small), C (medium) and D (large). Batteries made from these cells are usually rated at 6 V or 9 V.

Zinc-carbon cells have a fairly long shelf life. The zinc forms the outer case and is the negative electrode. A carbon rod serves as the positive electrode. The electrolyte is a paste of manganese dioxide and carbon. A zinc-carbon cell is inexpensive, and is usable at moderate temperatures and in applications where the current drain is moderate to high. This type of cell does not work well in extremely cold environments.

Alkaline cells have granular zinc for the negative electrode, potassium hydroxide as the electrolyte, and a polarizer as the positive electrode. An alkaline cell can work at lower temperatures than a zinc-carbon cell. It also lasts longer when used to operate common electronic devices, and is therefore preferred for use in transistor radios, calculators, and portable cassette players. Its shelf life is longer than that of a zinc-carbon cell.

Transistor batteries are small, 9-V, box-shaped batteries with clip-on connectors on top. They consist of six tiny zinc-carbon or alkaline cells in series. Each of the cells supplies 1.5 V. This type of battery is used in low-current devices such as wireless remote-control boxes, electronic calculators, and smoke detectors.

Lantern batteries are rather massive and can deliver a fair amount of current. One type has spring contacts on the top. The other type has thumbscrew terminals. Besides keeping an incandescent bulb lit for awhile, this type of battery, usually rated at 6 V and consisting of four zinc-carbon or alkaline cells, can provide enough energy to operate a low-power communications radio.

Silver-oxide cells have button-like shapes, and can fit inside a wristwatch. They come in various sizes and thicknesses, all with similar appearance. A silver-oxide cell supplies 1.5 V, and offers excellent energy storage capacity considering its light weight. It also has a nearly flat discharge curve. Silver-oxide cells can be stacked to make batteries about the size of an AAA cylindrical cell.

Mercury cells, also called *mercuric oxide cells*, have advantages similar to silver-oxide cells. They are manufactured in the same general form. The main difference, often not of significance, is a somewhat lower voltage per cell: 1.35 V. There has been a decrease in the popularity of mercury cells and batteries in recent years, because mercury is toxic and is not easily disposed of.

Lithium cells supply 1.5 V to 3.5 V, depending on the chemistry used. These cells, like their silver-oxide cousins, can be stacked to make batteries. Lithium cells and batteries have superior shelf life, and they can last for years in very-low-current applications. This type of cell or battery provides excellent energy capacity per unit volume.

Lead-acid cells and batteries have a solution or paste of sulfuric acid, along with a lead electrode (negative) and a lead-dioxide electrode (positive). Paste-type lead-acid batteries can be used in consumer devices that require moderate current, such as laptop computers and portable camcorders.

NICKEL-BASED CELLS AND BATTERIES

Nickel-cadmium (NICAD) cells and *nickel-metal-hydride (NiMH) cells* are made in several types. *Cylindrical cells* look like dry cells. *Button cells* are used in cameras, watches, memory backup applications, and other places where miniaturization is important. *Flooded cells* are used in heavy-duty applications, and can have a storage capacity of as much as 1000 Ah. *Spacecraft cells* are made in packages that can withstand extraterrestrial temperatures and pressures.

NICAD batteries and *NiMH batteries* are available in packs of cells that can be plugged into equipment to form part of the case for a device. An example is the battery pack for a handheld radio transceiver.

Nickel-based cells and batteries should never be left connected to a load after the current drops to zero. This can cause the polarity of a cell, or of one or more cells in a battery, to reverse. Once this happens, the cell or battery will no longer be usable. When a nickel-based cell or battery is "dead," it should be recharged as soon as possible.

NiMH cells and batteries are preferable to the older NICAD types. This is because cadmium, found in NICAD units, is toxic. In this respect, cadmium resembles mercury. Its disposal can present an environmental problem if not carried out properly. In fact, laws in many regions provide for strict procedures that must be followed when mercury or NICAD cells and batteries are discarded. Violations can be punished by high fines.

PROBLEM 4-8
Suppose you want to use a battery to back up the memory contents of a microcomputer-controlled device for months at a time during which the device is not used. The current required for this application is very low. You want the battery to have low mass and small volume, and to require no maintenance. What type of battery is good for this purpose?

SOLUTION 4-8
A lithium battery would work well in this situation. In fact, lithium batteries can be found in consumer devices of this sort, from radio transceivers to portable computers and cell phones.

Specialized Power Supplies

Some electronic systems employ specialized power sources, including *power inverters*, *uninterruptible power supplies*, and *solar-electric energy systems*.

POWER INVERTER

A power inverter, sometimes called a *chopper power supply*, is a circuit that delivers high-voltage AC from a low-voltage DC source. The input is typically 12 V DC, and the output is usually 117 V rms AC.

A simplified block diagram of a power inverter is shown in Fig. 4-17. The *chopper* consists of a low-frequency oscillator that opens and closes a high-current switching transistor. This interrupts the battery current, producing pulsating DC. The transformer converts the pulsating DC to AC, and also steps up the voltage.

The output of a low-cost power inverter is not necessarily a sine wave, but instead often resembles a sawtooth or square wave, and the frequency might be somewhat higher or lower than the standard 60 Hz. More sophisticated (and expensive) inverters produce fairly good sine-wave output, and have a frequency close to 60 Hz. High-end power inverters are preferred for use with sensitive equipment such as computers.

If the transformer shown in the block diagram is followed by a rectifier and filter, the device becomes a *DC transformer*, also called a *DC-to-DC*

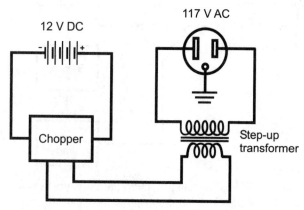

Fig. 4-17.   A power inverter converts low-voltage DC into high-voltage AC.

converter. Such a circuit can provide hundreds of volts DC when a 12 V battery is used as the power source.

UNINTERRUPTIBLE POWER SUPPLIES

When a piece of electronic equipment is operated from utility power, there is a possibility of a system malfunction or failure resulting from a *blackout*, *brownout*, *interruption*, or *dip*. A blackout is a complete loss of power for an extended period. A brownout is a significantly reduced voltage for an extended period. An interruption is a complete loss of power for a brief period. A dip is a significantly reduced voltage for a few moments.

To prevent power "glitches" from causing big trouble such as computer data loss, an *uninterruptible power supply (UPS)* can be used. Figure 4-18 is a block diagram of a UPS. Under normal conditions, the equipment gets its power through the transformer and regulator. The regulator eliminates

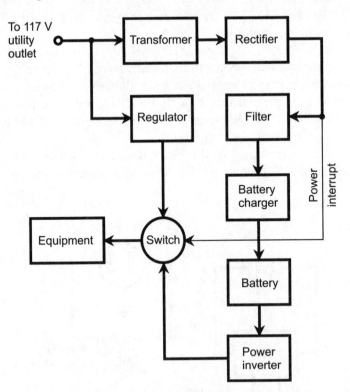

Fig. 4-18. An uninterruptible power supply (UPS) prevents system failure in case of utility power interruptions or irregularities.

transients, surges, and dips in the utility power. A lead-acid battery is kept charged by a small current through the rectifier and filter. If a utility power anomaly occurs, an *interrupt signal* causes the switch to disconnect the equipment from the regulator and connect it to the power inverter, which converts the battery DC output to AC. When utility power returns to normal, the switch disconnects the equipment from the battery and reconnects it to the regulator.

If power to a computer fails and you have a UPS, save all your work immediately on the hard drive, and also on an external medium such as a diskette if possible. Then switch the entire system, including the UPS, off until utility power returns.

SOLAR-ELECTRIC POWER SUPPLIES

Electric energy can be obtained directly from sunlight by means of *solar cells*, also known as *photovoltaic (PV) cells*. These devices, which are specialized semiconductor diodes, produce a few milliwatts of power for each square centimeter of surface area exposed to bright sunlight. Solar cells produce DC, while most household appliances require AC at 117 V and 60 Hz. Most solar-electric energy systems intended for general home and business use must therefore employ power inverters.

In some solar-electric power supplies, inverters are not necessary. For example, if you plan to operate a notebook computer from a battery and to keep the battery charged using a solar panel, you will not need a power inverter. But the charging circuit must be carefully designed so the battery is not damaged during the charging process.

There are two types of solar electric energy systems capable of powering homes and small businesses: the *stand-alone system* and *the interactive system*.

A stand-alone system uses banks of rechargeable batteries, such as the lead-acid type, to store electric energy as it is supplied by PV cells during hours of bright sunshine. The energy is released by the rechargeable batteries at night or in gloomy daytime weather. This system is independent of the electric utility, or *power grid*.

An interactive system is connected into the power grid. This type of system does not normally use storage batteries. Excess energy, if any, is sold to utility companies during times of daylight and minimum usage. Energy is bought from the utility at night, during gloomy daytime weather, or during times of heavy usage.

Quiz

Refer to the text in this chapter if necessary. Answers are in the back of the book.

1. Imagine that you want to obtain 480 V rms AC across the secondary winding of a power transformer that is to be connected to a source of 120 V rms AC. What is the primary-to-secondary turns ratio that this transformer must have?
    (a) 1:2
    (b) 1:4
    (c) 2:1
    (d) 4:1

2. Suppose, in the situation described in Question 1, you want to obtain DC and decide to use a half-wave rectifier circuit. Which of the following diode PIV ratings will be large enough (or more than large enough) for this?
    (a) 480 PIV
    (b) 960 PIV
    (c) 1440 PIV
    (d) None of the above

3. Suppose, in the situation described in Question 1, you want to obtain DC and decide to use a full-wave bridge rectifier circuit. Which of the following diode PIV ratings will be large enough (or more than large enough) for this?
    (a) 480 PIV
    (b) 960 PIV
    (c) 1440 PIV
    (d) None of the above

4. An automotive battery supplies 12 V DC and is rated at 60 Ah when fully charged. If it is charged up and then connected to a load that draws 500 mA of current, how long will this battery last before it needs recharging?
    (a) 30 hours
    (b) 60 hours
    (c) 120 hours
    (d) None of the above

5. An automotive battery supplies 12 V DC and is rated at 60 Ah when fully charged. If it is charged up and then connected to a load that

has a resistance of 48 Ω, how long will this battery last before it needs recharging?
(a) 30 hours
(b) 60 hours
(c) 120 hours
(d) None of the above

6. If a power transformer with a primary-to-secondary turns ratio of 1 : 3 is connected to a source of 120 V rms AC at 60 Hz, the frequency of the output, in hertz, is
(a) 20
(b) 60
(c) 180
(d) None of the above

7. Which of the following protects a power supply from damage caused by excessive current demand, by opening the circuit in case of malfunction?
(a) A power inverter.
(b) A solar cell.
(c) A rectifier diode.
(d) A circuit breaker.

8. If a power transformer with a primary-to-secondary turns ratio of 2 : 1 is connected to a source of 220 V rms AC, then the output is
(a) 55 V rms AC.
(b) 110 V rms AC.
(c) 440 V rms AC.
(d) 880 V rms AC.

9. A stand-alone PV system can operate
(a) independently of the utility power grid.
(b) by discharging its batteries into the utility power grid.
(c) only at night or on a gloomy day.
(d) by selling power to the utility company.

10. If you want to operate a common household floor lamp from a 12-V automotive battery, you will need to use
(a) an uninterruptible power supply.
(b) a full-wave bridge rectifier.
(c) a chopper power supply.
(d) a step-down transformer.

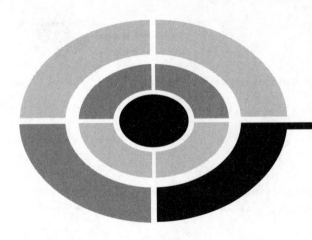

Test: Part One

Do not refer to the text when taking this test. You may draw diagrams or use a calculator if necessary. A good score is at least 30 correct. Answers are in the back of the book. It's best to have a friend check your score the first time, so you won't memorize the answers if you want to take the test again.

1. Which of the following conditions produces a high risk of data loss in magnetic tape?
   (a) Exposure to light.
   (b) Exposure to heat.
   (c) Freezing temperatures.
   (d) Low air pressure.
   (e) All of the above

2. In an AC sine wave, the frequency in hertz is equal to the
   (a) period in seconds.
   (b) reciprocal of the period in seconds.
   (c) peak voltage in volts.
   (d) peak-to-peak voltage in volts.
   (e) average voltage in volts.

3. What is the absolute-value impedance, Z, represented by the vectors
    shown in Fig. Test1-1?
    (a) $0\,\Omega$
    (b) $30\,\Omega$
    (c) $40\,\Omega$
    (d) $50\,\Omega$
    (e) It cannot be determined without more information.

4. The vector pointing straight up in Fig. Test1-1 represents
    (a) a pure resistance.
    (b) a pure inductive reactance.
    (c) a pure capacitive reactance.
    (d) a mixture of resistance and inductive reactance.
    (e) a mixture of resistance and capacitive reactance.

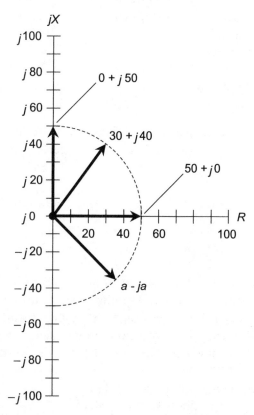

Fig. Test1-1.   Illustration for Part One Test Questions 3 through 6.

5. Consider the vector $a - ja$ in Fig. Test1-1. What is the approximate value of the constant, a?
    (a) There is no way to know from this information.
    (b) 25.00
    (c) 35.36
    (d) 50.00
    (e) 70.71

6. In Fig. Test1-1, which, if any, of the vectors represents a pure resistance?
    (a) The one pointing straight up.
    (b) The one pointing up and to the right.
    (c) The one pointing directly to the right.
    (d) The one pointing down and to the right.
    (e) None of them.

7. An AC wave with a practically instantaneous rise time and a gradual, straight-line decay time is known as
    (a) a sine wave.
    (b) a square wave.
    (c) a sawtooth wave.
    (d) a rectangular wave.
    (e) an irregular wave.

8. Suppose a transformer can reliably provide up to 40 A at 20 V rms AC output. What is the volt-ampere (VA) capacity of this transformer?
    (a) 2 VA
    (b) 20 VA
    (c) 40 VA
    (d) 60 VA
    (e) 800 VA

9. Consider a system described as follows: "Surplus energy generated by this system can be sold to the electric utility company during times of daylight and low usage. At night, during gloomy daytime weather, or during times of heavy usage, energy can be bought from the utility to supplement the output from this system." What is this system?
    (a) An electrochemical power system.
    (b) An interactive photovoltaic system.
    (c) A stand-alone photovoltaic system.

(d) A DC voltage-doubler power system.

(e) A power-inverting system.

10. The most expensive component in a power supply designed to provide DC output from utility AC input is usually the

(a) filter capacitor.

(b) diode.

(c) transformer.

(d) transient suppressor.

(e) circuit breaker or fuse.

11. Fill in the blank to make the following sentence true: "Current is the _____ electrical charge carriers flow past a specific point in a circuit."

(a) direction in which

(b) mass with which

(c) acceleration with which

(d) rate at which

(e) spin with which

12. What is the absolute value of the complex impedance $5 - j12$?

(a) $5\,\Omega$

(b) $12\,\Omega$

(c) $-12\,\Omega$

(d) $13\,\Omega$

(e) It cannot be determined without more information.

13. Fill in the blank to make the following sentence true: "If a series resistance-inductance-capacitance (RLC) circuit has inductive and capacitive reactances that cancel each other out at a particular frequency, then the circuit is _____ at that frequency."

(a) reactive

(b) resonant

(c) dissonant

(d) inoperative

(e) an open circuit

14. Suppose the peak voltage of a sine wave is approximately 1 kV, and the DC component is zero. What is the approximate peak-to-peak voltage?

(a) It cannot be determined without more information.

(b) 2000 V

(c) 1000 V

(d) 707 V

(e) 500 V

15. Suppose the peak voltage of a sine wave is approximately 1 kV, and the DC component is zero. What is the approximate root-mean-square (rms) voltage?

(a) It cannot be determined without more information.

(b) 2000 V

(c) 1000 V

(d) 707 V

(e) 500 V

16. How many "flashlight cells" must be connected in parallel to obtain 13.5 V DC?

(a) Two

(b) Four

(c) Six

(d) Eight

(e) It is impossible to get 13.5 V DC by connecting any number of "flashlight cells" in parallel.

17. Suppose a transformer has a primary-to-secondary turns ratio of 9 : 1. If 117 V rms AC is applied to the input, what rms AC voltage appears at the output?

(a) 39 V rms AC

(b) 13 V rms AC

(c) 1.44 V rms AC

(d) 351 V rms AC

(e) None; the output will be DC, not AC.

18. Examine Fig. Test1-2. Suppose the battery supplies a voltage $E = 6$ V DC, and the current $I = 2$ A. What is the resistance R?

(a) 3 Ω

(b) 12 Ω

(c) 1/3 Ω

(d) 8 Ω

(e) It cannot be determined without more information.

19. In Fig. Test1-2, suppose the resistance $R = 120\,\Omega$ and the current $I = 100$ mA. What is the battery voltage?

(a) 12,000 V

(b) 1200 V

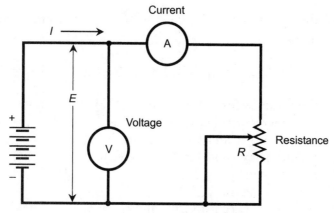

Fig. Test1-2. Illustration for Part One Test Questions 18 through 20.

(c) 220 V

(d) 12 V

(e) It cannot be determined without more information.

20. In Fig. Test1-2, suppose the voltage across the resistor is 18 V. What
is the current flowing through ammeter A?

(a) 0.18 A

(b) 1.8 A

(c) 18 A

(d) 180 A

(e) It cannot be determined without more information.

21. Imagine an inductor whose value in microhenrys (μH) is L, and
that carries a signal whose frequency in megahertz (MHz) is f. The
inductive reactance of this component, X_L, is given by the following
formula:

$$X_L = 2\pi f L$$

Based on this information, we can say for certain that as the
frequency increases, the reactance of a 10 μH inductor

(a) becomes larger and larger positively.

(b) becomes larger and larger negatively.

(c) stays the same.

(d) approaches zero.

(e) alternates between positive and negative.

22. Imagine a capacitor whose value in microfarads (μF) is C, and
that carries a signal whose frequency in megahertz (MHz) is f. The

capacitive reactance of this component, X_C, is given by the following formula:

$$X_C = -1/(2\pi f C)$$

Based on this information, we can say for certain that as the frequency increases, the reactance of an 0.001 μF capacitor
(a) becomes larger and larger positively.
(b) becomes larger and larger negatively.
(c) stays the same.
(d) approaches zero.
(e) alternates between positive and negative.

23. Long-distance utility power transmission lines operate at high voltages in order to
(a) maximize the EMF radiation.
(b) produce powerful magnetic fields.
(c) repel lightning.
(d) minimize the power lost as heat in the wires.
(e) make the rectified current easy to filter.

24. Fill in the blank to make the following sentence true: "A radio-frequency (RF) transmission line's _____ depends on the cross-sectional physical dimensions of the line, and also on the nature of the material separating the conductors."
(a) inherent resistance
(b) voltage ratio
(c) resonant frequency
(d) wavelength limit
(e) characteristic impedance

25. The diodes in a power supply make it possible to
(a) change DC to AC.
(b) step up the AC output voltage.
(c) rectify the input.
(d) filter the output.
(e) keep the output from shorting out.

26. Which of the following is not a source of DC?
(a) A photovoltaic cell exposed to direct sunlight.
(b) A size C flashlight cell.
(c) A lantern battery.

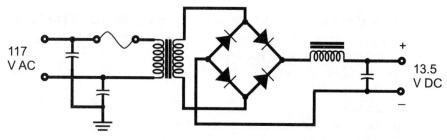

Fig. Test1-3.   Illustration for Part One Test Questions 27 through 30.

(d) The electricity at a common American household utility outlet.
(e) The output of a rectified and filtered power supply.

27. What is the circuit shown in Fig. Test1-3?
   (a) A device for testing diodes.
   (b) An electrochemical power system.
   (c) A power inverter.
   (d) A DC power supply.
   (e) An AC step-up transformer.

28. What are the functions of the capacitors to the left of the transformer in Fig. Test1-3?
   (a) They convert the incoming AC into DC.
   (b) They rectify the incoming AC.
   (c) They suppress transients in the incoming AC.
   (d) They reduce the incoming AC voltage to 13.5 V.
   (e) None of the above

29. What is the function of the capacitor at the far right in Fig. Test1-3?
   (a) It rectifies the AC into DC.
   (b) It reduces the DC output voltage.
   (c) It keeps the diodes from malfunctioning.
   (d) It keeps the output voltage from getting too low.
   (e) None of the above

30. What would happen if the coil to the right of the diodes in Fig. Test1-3 were to short out?
   (a) The output would be less well filtered.
   (b) The fuse would blow.
   (c) The output voltage would be greatly reduced.
   (d) The output would change from DC to AC.
   (e) The output polarity would be reversed.

31. Fill in the blank to make the following sentence true: "The intensity of the magnetic field surrounding a wire that carries DC _____ as the current in the wire increases."
    (a) remains constant
    (b) increases
    (c) decreases
    (d) alternately increases and decreases
    (e) fluctuates at random

32. In Fig. Test1-4, wave X and wave Y appear to be
    (a) in phase coincidence.
    (b) in phase opposition.
    (c) 90° out of phase.
    (d) 270° out of phase.
    (e) 360° out of phase.

33. In Fig. Test1-4, wave X and wave Y appear to have
    (a) identical frequencies.
    (b) opposite frequencies.
    (c) frequencies that differ by 90°.
    (d) frequencies that differ by 270°.
    (e) frequencies that differ by 360°.

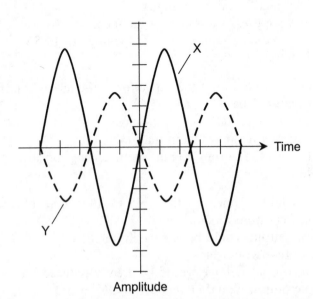

Fig. Test1-4.   Illustration for Part One Test Questions 32 through 35.

34. In Fig. Test1-4, suppose each horizontal division represents 0.001 second (or 1 millisecond). Based on this information, it appears that the frequency of wave Y is
    (a) impossible to determine without more information.
    (b) 1000 kHz.
    (c) 1 kHz.
    (d) 500 Hz.
    (e) 200 Hz.

35. In Fig. Test1-4, suppose each horizontal division represents 0.001 second (or 1 millisecond). Based on this information, it appears that the peak amplitude of wave X is
    (a) impossible to determine without more information.
    (b) 1000 kV.
    (c) 1 kV.
    (d) 500 V.
    (e) 200 V.

36. Suppose five resistors, each having a resistance of 47 Ω, are connected in series. What is the total resistance of the combination?
    (a) More information is needed to answer this question.
    (b) 4.7 Ω
    (c) 9.4 Ω
    (d) 47 Ω
    (e) 235 Ω

37. Suppose five resistors, each having a resistance of 47 Ω, are connected in parallel. What is the total resistance of the combination?
    (a) More information is needed to answer this question.
    (b) 4.7 Ω
    (c) 9.4 Ω
    (d) 47 Ω
    (e) 235 Ω

38. How many radians of phase are there in one-quarter of an AC cycle?
    (a) $\pi/4$
    (b) $\pi/2$
    (c) π
    (d) 2π
    (e) 4π

39. In the impedance $Z = 4 + j7$, the reactance is
    (a) resistive.
    (b) capacitive.

(c) inductive.

(d) complex.

(e) zero.

40. Which of the following properties is related to the ability of a material to concentrate magnetic flux?

(a) Resistance

(b) Conductance

(c) Permeability

(d) Amperage

(e) Charge-carrier capacity

Wired Electronics

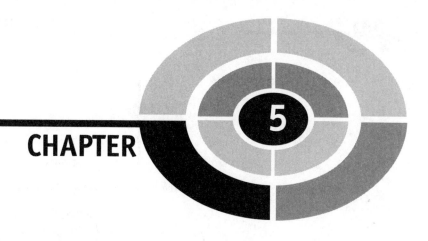

CHAPTER 5

Semiconductor Diodes

Most diodes are manufactured with materials that "sort of conduct" electricity. These substances are called *semiconductors*. There are two main categories: *N type*, in which the charge carriers are mainly electrons, and *P type*, in which the charge carriers are primarily *holes* (spaces in atoms where electrons, normally present, are missing).

The P-N Junction

When wafers of N type and P type semiconductor material are in direct physical contact, the result is a *P-N junction* with unique and useful properties. Figure 5-1 shows the schematic symbol for a semiconductor diode, which is a two-electrode device. The N type material is represented

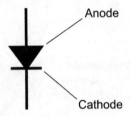

Fig. 5-1.   Anode and cathode symbology for a diode.

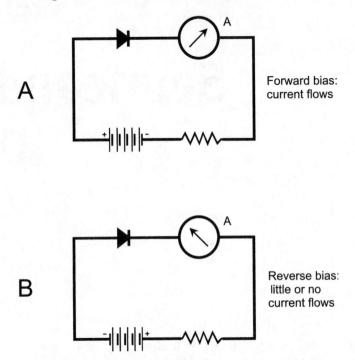

Fig. 5-2.   At A, forward bias results in a flow of current. At B, moderate reverse bias produces near zero current.

by the short, straight line in the symbol, and forms the *cathode*. The P type material is represented by the arrow, and forms the *anode*.

In a conventional diode, electrons flow easily in the direction opposite the arrow, and hardly at all in the direction that the arrow points (with an exception to be mentioned shortly). If a battery and a resistor are connected in series with the diode, current will flow if the negative terminal of the battery is connected to the cathode and the positive terminal is connected to the anode (Fig. 5-2A). This condition is called *forward bias*. Virtually no current will flow if the battery is reversed (Fig. 5-2B). This condition is called *reverse bias*.

FORWARD BREAKOVER VOLTAGE

It takes a certain minimum voltage for conduction to occur when a diode is forward-biased. This voltage is called the *forward breakover voltage*. Depending on the type of material, this voltage can range from about 0.3 V to 1 V. In a *silicon diode*, the most common type, it is 0.5 V. If the voltage across the junction is not at least as great as the forward breakover voltage, the diode will not conduct.

BIAS

When the N type material is negative with respect to the P type, a diode is forward biased and electrons flow easily from N to P. A forward-biased diode conducts well as long as the voltage is at least equal to the forward breakover voltage. The P-N junction in a forward-biased diode conducts electricity like a piece of silver or copper.

When the polarity is switched so the diode is reverse-biased, it conducts poorly under most circumstances. In the N type material, electrons are pulled towards the positive charge pole, away from the P-N junction. In the P type material, holes are pulled toward the negative charge pole, also away from the P-N junction. The electrons (in the N type material) and holes (in the P type) practically disappear in the vicinity of the junction. This impedes conduction, and the resulting *depletion region* behaves as a *dielectric* or electrical insulator. The P-N junction in a reverse-biased diode prevents the flow of electricity, like glass or dry paper.

JUNCTION CAPACITANCE

Under conditions of reverse bias, a P-N junction can act as a capacitor. The *varactor diode* is made with this property specifically in mind. The *junction capacitance* can be varied by changing the reverse-bias voltage, because this voltage affects the width of the depletion region. The greater the reverse voltage, the wider the depletion region gets, and the smaller the capacitance becomes.

AVALANCHE EFFECT

If a diode is reverse biased and the voltage is increased without limit, a point will be reached at which the P-N junction will suddenly begin to conduct.

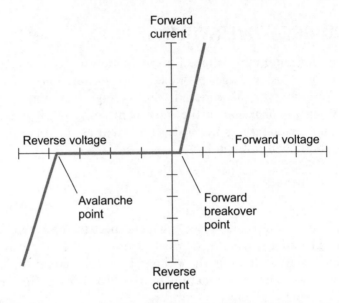

Fig. 5-3. Characteristic curve for a hypothetical diode, showing the forward breakover point and the avalanche point.

Its insulating properties will be overcome, and electrons will flow in large quantities. This is known as *avalanche effect*. The minimum voltage at which this occurs is called the *avalanche voltage*, and it varies among different kinds of diodes.

Figure 5-3 is a graph of the characteristic current-versus-voltage curve for a typical semiconductor diode, showing the *avalanche point*. The avalanche voltage is considerably greater than, and is of the opposite polarity from, the forward breakover voltage.

RECTIFICATION

A *rectifier diode* passes current in only one direction under normal operating conditions. This makes it useful for changing AC to DC. You learned all about this in Chapter 4. Avalanche effect is undesirable in diodes when they are used as rectifiers, because it degrades the efficiency of a power supply. The phenomenon can be prevented by choosing diodes with an avalanche voltage higher than the *peak inverse voltage* (PIV) produced by the power supply circuit. For extremely high-voltage power supplies, it is necessary to use series combinations of diodes to prevent avalanche effect.

ZENER DIODES

Most diodes have avalanche voltages that are much higher than the reverse bias ever gets. The value of the avalanche voltage depends on how a diode is manufactured. *Zener diodes* are made to have well-defined, constant avalanche voltages. This type of diode takes advantage of the avalanche effect, and is the basis for voltage regulation in some power supplies.

Suppose a certain Zener diode has an avalanche voltage, also called the *Zener voltage*, of 50 V. If a reverse bias is applied to the P-N junction, the diode acts as an open circuit below 50 V. When the voltage reaches 50 V, the diode starts to conduct. The more the reverse bias tries to increase, the more current flows through the P-N junction. This effectively prevents the reverse voltage from exceeding 50 V.

There are other ways to get voltage regulation besides the use of Zener diodes, but Zener diodes often provide the simplest and least expensive alternative. Zener diodes are available with a wide variety of voltage and power-handling ratings. Power supplies for solid-state equipment commonly employ Zener diode regulators.

THYRECTORS

A *thyrector* is a semiconductor device that is basically equivalent to two diodes connected in series, but with their polarities reversed. This component is used to protect equipment against the effects of transients in the voltage supply.

A pair of thyrectors can be connected between each of the AC power terminals of a piece of electronic apparatus and electrical ground (Fig. 5-4).

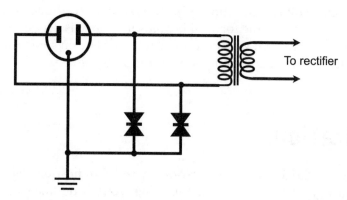

To rectifier

Fig. 5-4.   Thyrectors can suppress transients in an AC line.

The device normally acts as an open circuit. It does not conduct until the voltage reaches a certain value. If the voltage across the thyrector exceeds the critical level, even for a tiny fraction of a second, then it conducts. This in effect shorts transients out by keeping the peak voltage from exceeding the critical level, called the *clipping voltage*. This eliminates any adverse effect on the equipment.

Thyrectors are effective against most, but not all, transients. A severe spike, such as might be induced by the *electromagnetic pulse* (EMP) from a nearby lightning strike, can destroy a thyrector and the equipment with which it is used. In thunderstorms, the best way to protect a piece of equipment is to physically unplug its power supply from the wall outlet. This is sometimes laughed at as the "little old lady" approach – but it works.

PROBLEM 5-1
What is the difference between forward bias, reverse bias, and zero bias in a diode?

SOLUTION 5-1
In forward bias, the cathode (N type material) is at a negative potential (voltage) with respect to the anode (P type material). In reverse bias, the cathode is positive with respect to the anode. In zero bias, the two electrodes are at the same electrical potential.

PROBLEM 5-2
How does the avalanche voltage of a typical diode compare with the forward breakover voltage?

SOLUTION 5-2
The avalanche voltage is defined for reverse bias (cathode positive with respect to anode), while the forward breakover voltage is defined for forward bias (anode positive with respect to cathode). The avalanche voltage is usually much greater than the forward breakover voltage.

Signal Applications

Semiconductor diodes are used in various types of audio-frequency (AF) and radio-frequency (RF) circuits. The following sections outline some common applications of so-called *signal diodes*.

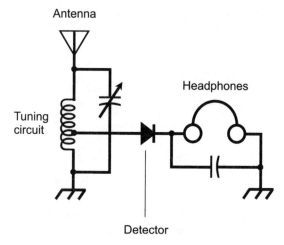

Fig. 5-5. A diode as an envelope detector in a crystal radio receiver.

ENVELOPE DETECTION

One of the earliest diodes, existing even before vacuum tubes, was a semiconductor. Known as a *cat whisker*, it consisted of a fine piece of wire in contact with a small piece of the mineral substance *galena*. It acted as a *detector*, the equivalent of a rectifier for small RF currents. When the cat whisker was connected in a circuit like that of Fig. 5-5, the result was a *crystal radio receiver* capable of picking up amplitude-modulated (AM) radio signals. Using an RF diode, this circuit can still be used today. It works without any external source of power other than the signal itself.

The diode acts to recover the audio from the radio signal. This is *envelope detection*. If the detector is to be effective, the diode must have certain characteristics. It should have low junction capacitance so that it works as a rectifier (not as a capacitor) at RF, passing current in one direction but not in the other. Some modern RF diodes are microscopic versions of the old cat whisker. These are known as *point-contact diodes*, and are designed to minimize the junction capacitance.

ELECTRONIC SWITCHING

The ability of diodes to conduct when forward biased, and to insulate when reverse biased, makes them useful for *electronic switching* in some applications. Diodes can switch signals much faster than any mechanical device.

One type of diode, made for use as an RF switch, has a special semiconductor layer sandwiched in between the P type and N type material. This

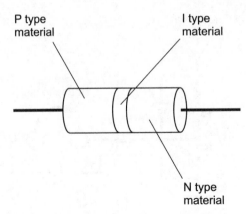

P type
material

I type
material

N type
material

Fig. 5-6. A PIN diode has a thin layer of intrinsic (I type) material between the P and N type
semiconductor layers.

layer, called an *intrinsic semiconductor*, reduces the junction capacitance so the
device can work at higher frequencies than an ordinary diode. The intrinsic
material is called *I type* material. A diode with an I type semiconductor layer
between the cathode and the anode is called a *PIN diode* (Fig. 5-6).

Direct-current bias, applied to one or more PIN diodes, allows RF
currents to be effectively channeled without using complicated relays and
cables. The electronic switch is closed when the diode is forward-biased, and
is open when the diode is reverse-biased. A PIN diode also makes a good RF
envelope detector because of its low junction capacitance.

FREQUENCY MULTIPLICATION

When fluctuating current passes through a diode, the shape of the output
wave is much different from the shape of the input wave. This is called
nonlinearity, and it results in an output signal rich in *harmonics* (signals
whose frequencies are whole-number multiples of the input signal frequency).
This effect makes diodes useful for *frequency multiplication*.

A simple frequency-multiplier circuit is shown in   Fig. 5-7. The output
inductance-capacitance (LC) circuit is tuned to the desired nth harmonic
frequency, nf_o, rather than to the input or *fundamental frequency*, f_o. For
example, if you want to multiply an input signal at 5 MHz by a factor of 3,
the output is tuned to $3 \times 5 = 15$ MHz.

For a diode to work as a frequency multiplier, it must be of a type that
would also work well as a detector at the same frequencies. This means that
the component should behave as a rectifier, and not as a capacitor. The PIN
diode is a good choice in this situation, although there are other types that

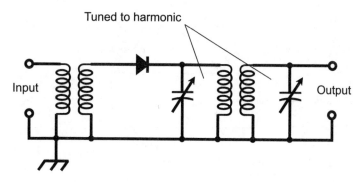

Fig. 5-7. A diode can be used in a frequency-multiplier circuit.

work even better. Power-supply rectifier diodes, in general, are not suitable for use as frequency multipliers at RF, because the junction capacitance is too high.

STEP-RECOVERY DIODES

A *step-recovery diode*, also called a *charge-storage diode* or *snap diode*, is a specialized device that works exceptionally well as a harmonic generator. When a signal is passed through a step-recovery diode, hundreds or even thousands of harmonic signals appear at the output.

When a step-recovery diode is forward-biased, current flows in the same manner as it would in any forward-biased P-N junction. But when reverse bias is applied, the conduction does not immediately stop. The step-recovery diode stores a large number of charge carriers while it is forward-biased, and it takes a short time for these charge carriers to be drained off. Current flows in the reverse direction until the charge carriers have been removed from the P-N junction, and then the current diminishes rapidly. The transition time between maximum current and zero current can be less than 0.1 nanosecond (0.0000000001 or 10^{-10} second)! This rapid change in current is responsible for the harmonic output.

HOT-CARRIER DIODES

The *hot-carrier diode* (HCD) is another device that is useful in RF circuits. This device can function at much higher frequencies than can most other types of semiconductor diodes because it's "quiet." It generates very little electrical noise (chaotic output that covers a broad range of frequencies). This feature is important in the design of radio receivers.

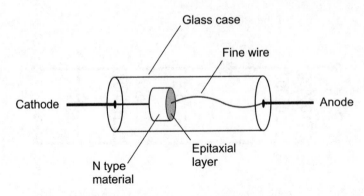

Fig. 5-8.   Construction of a hot-carrier diode.

The HCD has a high avalanche voltage. The reverse current is practically zero when the reverse bias is less than the avalanche voltage. The junction capacitance is low, because the diode consists of a tiny point contact (Fig. 5-8). The wire is plated with gold or platinum to minimize corrosion. An N-type silicon wafer is used as the semiconductor.

In an HCD, the silicon wafer is coated with a thin layer of aligned atoms called an *epitaxial layer*. This layer is produced by a lab process called *epitaxy*, in which crystals or layers of semiconductor are "grown" on a larger piece of material. It's a little like the accumulation of frost on your car windows when the temperature is below freezing and the humidity is high, but it's done under carefully controlled conditions.

SIGNAL MIXING

A *mixer* is a circuit that combines two signals of different frequencies to produce a third signal, the frequency of which is the sum or the difference of the input frequencies. One of the input signals is the one to be received, which comes into the mixer from preceding amplifiers. The other input signal comes from a special circuit called a *local oscillator* (LO), whose frequency can be precisely controlled.

Suppose there are two signals with frequencies f_1 and f_2, where f_1 is the lower frequency and f_2 is the higher frequency. If these signals are combined in a nonlinear device such as a diode, new signals result. One of the output signals occurs at the difference frequency $f_2 - f_1$, and the other output signal occurs at the sum frequency $f_2 + f_1$. The output signals are called *mixing products*, and the output frequencies are known as *beat frequencies*.

Mixing is used in radio and television receivers and transmitters. The technique is useful when it is necessary to convert input signals, which can occur over a wide range of frequencies, to a signal that stays at the same frequency all the time. A signal that never changes frequency is much easier to process than one whose frequency changes.

SINGLE BALANCED MIXER

A *single balanced mixer* is a mixer circuit that can be built at low cost. The main disadvantage of this circuit is the fact that the *input port* and the *output port* are not completely isolated, meaning that some of either input signal leaks through to the output. If good isolation is needed, a *double balanced mixer* is preferable.

A single-balanced-mixer circuit is shown in Fig. 5-9. The diodes are of the hot-carrier type. This circuit works at frequencies up to several gigahertz. Some signal loss occurs, but this loss can be overcome by an amplifier following the mixer.

DOUBLE BALANCED MIXER

The input signals in a double balanced mixer (Fig. 5-10) do not leak through to the output nearly as much as they do in the single balanced mixer. The output ports and input ports are isolated. The coupling (extent of signal transfer or interaction) between the input oscillators is negligible, and the coupling between the inputs and the output is negligible.

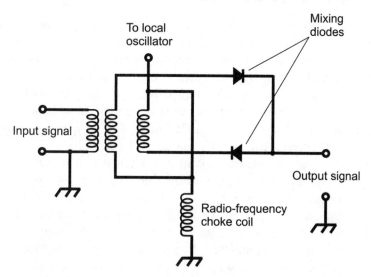

Fig. 5-9. A single balanced mixer using two hot-carrier diodes.

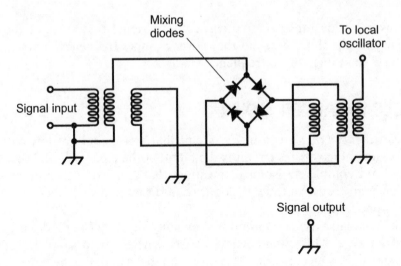

Fig. 5-10. A double balanced mixer using four hot-carrier diodes.

As in the single balanced mixer, hot-carrier diodes are used in this circuit. They can handle large signal amplitudes without distortion, and they generate very little noise. This circuit, like the single balanced mixer, has some conversion loss that can be overcome by an amplifier following the mixer.

AMPLITUDE LIMITING

The forward breakover voltage of a diode made from germanium is about 0.3 V. For a diode made from silicon, it is about 0.6 V. Remember that a diode does not conduct until the forward bias voltage is at least as great as the forward breakover voltage. There is an interesting corollary to this: a diode always conducts when the forward bias exceeds the breakover value, and the voltage across the diode is constant under these conditions: 0.3 V for germanium and 0.6 V for silicon. Regardless of how much or how little current flows, the voltage across the diode stays the same.

This property can be used to advantage when it is necessary to limit, or *clip*, the peak voltage of a signal, as shown in Fig. 5-11. When two identical diodes are connected in *reverse parallel* (that is, in parallel with opposing polarities, as shown in drawing A), the maximum peak amplitude is limited to the forward breakover voltage of the diodes. The input and output waveforms of a clipped signal are illustrated in drawing B. This scheme is sometimes used in the audio stages of communications receivers to prevent "blasting" when a strong signal comes in.

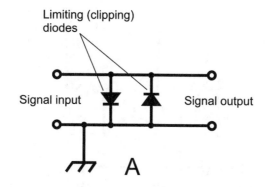

Limiting (clipping) diodes

Signal input

Signal output

A

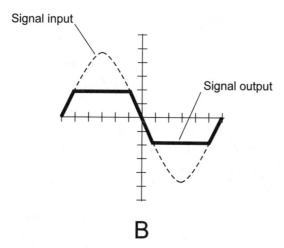

Signal input

Signal output

B

Fig. 5-11. At A, a limiter circuit using two diodes. At B, clipping of a signal passing through the limiter.

The chief disadvantage of the diode limiter circuit is that it introduces distortion when limiting is taking place. This is not necessarily a problem with digital data, or with analog signals that rarely reach the limiting voltage. But for analog voice or video signals with amplitude peaks that rise past the limiting voltage, it degrades the fidelity. In the worst cases, it can render a voice unintelligible or an image mutilated.

NOISE LIMITING

In a radio receiver, a *noise limiter* can consist of a pair of diodes connected in reverse parallel with variable reverse bias for control of the clipping

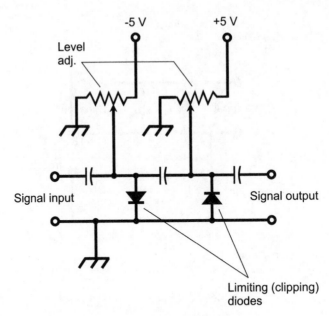

Fig. 5-12. A variable-threshold limiter circuit.

level (Fig. 5-12). The bias is adjusted using a *dual potentiometer*, which is a pair of identical variable resistors connected to a single rotatable shaft.

A limiting circuit such as the one shown in Fig. 5-12 makes it possible to set the peak amplitude at which everything passing through the circuit is clipped, from a minimum equal to the forward breakover voltage of the diodes to a maximum equal to the forward breakover voltage plus the voltage of the power supply (in this case ±5 V). When there is a lot of noise being received, in addition to the desired signal, the dual potentiometer is adjusted until clipping occurs at the signal amplitude. Noise pulses then cannot exceed the signal amplitude. This makes it possible to receive a signal that would otherwise be drowned out by noise.

FREQUENCY CONTROL

When a diode is reverse-biased, the depletion region at the P-N junction has dielectric properties that make the diode act like a capacitor. The width of this zone depends on several factors, including the reverse-bias voltage. As long as the reverse bias is less than the avalanche voltage, the width of the depletion region can be changed by varying the bias. This causes a change in the capacitance of the junction. The capacitance, which is on the order of a few picofarads, varies inversely with the square root of the reverse bias voltage.

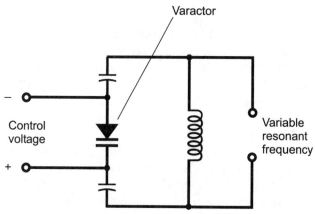

Fig. 5-13. A variable inductance-capacitance (*LC*) tuned circuit using a varactor.

Some diodes are manufactured especially for use as variable capacitors. These are *varactor diodes*, also called *varactors* or *varicaps*. They are made from silicon, or from a compound called *gallium arsenide*.

The varactor diode is used in a *voltage-controlled oscillator* (VCO). A parallel voltage-tuned inductor/capacitor (LC) circuit using a varactor is shown in Fig. 5-13. As the voltage across the diode varies, so does its capacitance; this affects the resonant frequency of the LC circuit, and therefore the frequency at which the oscillator produces a signal. The fixed capacitors, whose values are large compared with that of the varactor, prevent the coil from short-circuiting the control voltage across the varactor.

The symbol for a varactor diode has two lines, rather than one, on the cathode side. This emphasizes the fact that the capacitive effect of the depletion region is utilized, rather than the "one-way-current" property of the P-N junction.

PROBLEM 5-3
Suppose two signals, one having an input frequency of 3.500 MHz and the other having an input frequency of 12.500 MHz, are applied to the input of a mixer. What are the output signal frequencies?

SOLUTION 5-3
The output frequencies are equal to the sum and difference of the input signal frequencies. In this example, let $f_1 = 3.500$ MHz and $f_2 = 12.500$ MHz. Then:

$$f_2 - f_1 = 12.500 - 3.500 = 9.000 \text{ MHz}$$
$$f_2 + f_1 = 12.500 + 3.500 = 16.000 \text{ MHz}$$

PROBLEM 5-4

Suppose the difference output of a mixer is a constant 9.000 MHz. Imagine that you want to design a mixer circuit that will allow the reception of signals in the range 3.500 MHz to 4.000 MHz. What should be the range of frequencies for the local oscillator (LO), so a difference-frequency signal is produced at a constant 9.000 MHz?

SOLUTION 5-4

You will need to have two signals, f_2 and f_1, at the input, such that $f_2 - f_1 = 9.000$. Let f_1 represent the frequency of the signal to be received. It can range from 3.500 MHz to 4.000 MHz. The LO frequency, f_2, must always be 9.000 MHz greater than the received signal frequency in order to produce a difference output of 9.000 MHz. The tunable range of the LO should therefore be from $9.000 + 3.500$ to $9.000 + 4.000$ MHz, or 12.500 to 13.000 MHz.

Oscillation and Amplification

Under some conditions, diodes can generate or amplify radio signals at ultrahigh and microwave frequencies. These frequencies are generally greater than 300 MHz. Certain diodes are specifically manufactured for this purpose.

GUNN DIODES

A *Gunn diode* can produce up to 1 W of RF power output, but more commonly it works at levels of about 0.1 W. Gunn diodes are usually made from gallium arsenide (GaAs). The device oscillates because of the so-called *Gunn effect*, named after J. Gunn of International Business Machines (IBM) who observed it in the 1960s.

A Gunn diode does not function like a rectifier, detector, modulator, mixer, or clipper. Instead, oscillation takes place as a result of a property called *negative resistance*. This term, which at first seems like an oxymoron, means that over a certain portion of the diode's characteristic curve, the current decreases as the voltage increases. This property produces instability in the device under certain conditions, leading to the generation of RF signals.

Gunn-diode oscillators can be tuned using varactor diodes. A Gunn-diode oscillator, connected directly to a microwave *horn antenna*, is known as a *Gunnplexer*. These devices are popular with experimenters, especially amateur ("ham") radio operators.

IMPATT DIODES

The acronym *IMPATT* (pronounced "IM-pat") comes from the words *impact avalanche transit time*. This effect is similar to negative resistance. An IMPATT diode is a microwave oscillating device like a Gunn diode, except that it uses silicon rather than gallium arsenide.

An IMPATT diode can be used as an amplifier for a microwave transmitter that employs a Gunn-diode oscillator. As an oscillator, an IMPATT diode produces about the same amount of output power, at comparable frequencies, as a Gunn diode.

TUNNEL DIODES

Another type of diode that can oscillate at microwave frequencies is the *tunnel diode*, also known as the *Esaki diode*. It produces only a tiny amount of power, but it can be used as the LO in a microwave radio receiver.

Tunnel diodes work well as amplifiers in microwave receivers because they generate very little noise. This is especially true of gallium arsenide devices.

Diodes and Radiant Energy

Some semiconductor diodes emit infrared (IR) or visible light when a current passes through the P-N junction in a forward direction. This phenomenon, called *photoemission*, occurs as electrons fall from higher to lower energy states within atoms. Diodes can respond to incoming IR or visible light as well. Some *photosensitive diodes* have variable resistance that depends on the illumination intensity. Others actually generate DC voltages in the presence of IR or visible light.

LEDs AND IREDs

The most common color for a *light emitting diode* (LED) is bright red. An *infrared emitting diode* (IRED) produces wavelengths too long to see. The intensity of energy emission from an LED or IRED depends to some extent on the forward current. As the current rises, the brightness increases up to a certain point. If the current continues to rise, no further increase in brilliance takes place. The LED or IRED is then said to be in a state of *saturation*.

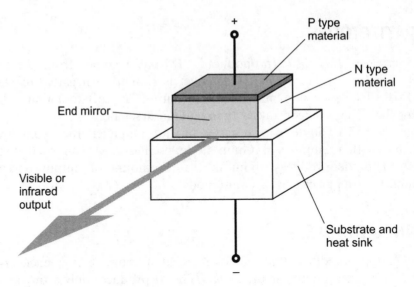
Fig. 5-14. Simplified view of the construction of a laser diode.

LASER DIODES

The *laser diode* is a form of LED or IRED with a relatively large, flat P-N junction on a metal or silicon *substrate*. The junction emits coherent waves, provided the applied current is sufficient. In coherent radiation, all waves are of the same frequency, and they are all in phase. If the current is below a certain level, the laser diode behaves like an ordinary LED or IRED, but when the threshold current is reached, the charge carriers recombine in such a manner that *lasing* (the effect that produces coherent radiation) occurs. The construction of a typical laser diode is shown in Fig. 5-14.

Laser diodes are used in optical communications systems, especially when the beams must travel considerable distances through the air or through space. Coherent light or IR suffers less attenuation than ordinary light or IR as it traverses long distances. This is because it can be focused into a narrow beam that does not diverge with distance as rapidly as a beam of conventional light or IR.

COMMUNICATIONS

Both LEDs and IREDs are useful in communications, because their intensity can be modulated to carry information. When the current through the device is sufficient to produce output, but not enough to cause saturation, the LED or IRED output follows along with rapid current changes. Voices, music, video, and digital signals can be conveyed over visible or IR beams in this way. Many data communications systems make use of modulated light or IR, transmitted through clear fibers. This is known as *fiberoptic technology*.

SILICON PHOTODIODES

A silicon diode, housed in a transparent case and constructed in such a way that visible light can strike the barrier between the P type and N type materials, forms a *photodiode*. A reverse DC bias is applied to the device. When visible light, IR, or ultraviolet (UV) falls on the junction, current flows. The current is proportional to the intensity of the incoming radiation, within certain limits.

When energy of varying intensity strikes the P-N junction of a reverse-biased silicon photodiode, the output current follows the fluctuations. This makes silicon photodiodes useful for receiving modulated-light signals, of the kind used in fiberoptic systems.

OPTOISOLATORS

An LED or IRED and a photodiode can be combined in a single package to form an *optoisolator*, also known as an *optical coupler*. With an electrical-signal input, the LED or IRED generates a modulated visible or IR beam and sends it over a small, clear gap to the photodiode, which converts the visible or IR energy back into an electrical signal.

An ongoing bugaboo for engineers is the fact that, when a signal is electrically coupled from one circuit to another, the two stages interact. For example, an oscillator might work fine all by itself, but when it is connected to an amplifier in an attempt to get a stronger signal, the oscillator fails because of the load imposed on it by the amplifier. This can lead to all kinds of problems. Optoisolators overcome this effect, because the coupling is not done electrically. If the electrical input impedance of the second circuit changes, the impedance that the first circuit "sees" is not affected.

PHOTOVOLTAIC CELLS

A silicon diode, with no bias voltage applied, generates DC all by itself if sufficient visible, IR, or UV energy strikes its P-N junction. This is known as the *photovoltaic effect* and is the principle by which solar cells work.

Photovoltaic (PV) cells have large P-N junction surface area (Fig. 5-15). This maximizes the amount of energy that can strike the junction. A single

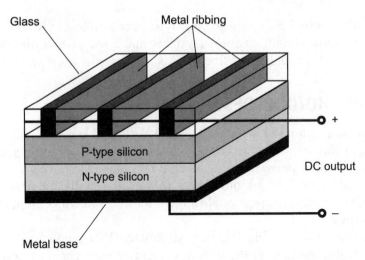

Glass   Metal ribbing

P-type silicon

N-type silicon   DC output

+

−

Metal base

Fig. 5-15.   Simplified view of the construction of a photovoltaic (PV) cell.

silicon PV cell produces approximately 0.6 V DC in direct sunlight with no load. The amount of current that it can deliver depends on the surface area of the junction.

Silicon PV cells are connected in series-parallel combinations to provide solar power for solid-state electronic devices such as portable radios. A large assembly of such cells constitutes a *solar panel*. The DC voltages of the cells add when they are connected in series. A typical *solar battery* supplies 6, 9, or 12 V DC.

PROBLEM 5-5
Why does a silicon PV cell produce only about 0.6 V DC?

SOLUTION 5-5
This is the case because 0.6 V is the forward breakover voltage of a silicon P-N junction. The output of a semiconductor PV cell appears with a polarity that is the same as forward bias. This tends to limit the voltage that can exist across the junction, in a manner similar to the way a diode limiter works.

PROBLEM 5-6
Suppose a single silicon PV cell can produce 100 mA of current in direct sunlight. How many of these would be necessary, if connected in parallel, to produce 500 mA?

SOLUTION 5-6
Five PV cells in parallel would serve this purpose. Currents in parallel add directly.

Quiz

Refer to the text in this chapter if necessary. Answers are in the back of the book.

1. A diode that is used as an envelope detector should have
    (a) high forward breakover voltage.
    (b) high avalanche voltage.
    (c) large negative resistance.
    (d) small junction capacitance.

2. A varactor can be used for the purpose of
    (a) tuning a Gunn-diode oscillator.
    (b) photoemission.
    (c) generation of electricity using photovoltaic effect.
    (d) audio amplification.

3. The forward breakover voltage of a signal diode can be exploited in the design of
    (a) a rectifier circuit.
    (b) a solar energy system.
    (c) a noise limiter.
    (d) an oscillator.

4. The maximum amount of current that a silicon PV cell can deliver in direct sunlight depends on
    (a) the polarity of the bias.
    (b) the current in the reverse direction.
    (c) the forward breakover voltage.
    (d) None of the above

5. As electrons in the atoms of a semiconductor material fall from higher energy levels to lower levels,
    (a) the junction capacitance increases.
    (b) the device stops conducting.
    (c) photoemission can occur.
    (d) the forward breakover voltage fluctuates.

6. In a conventional diode, once the reverse bias across the P-N junction exceeds the avalanche voltage, a further increase in reverse bias causes
    (a) an increase in the current through the junction.
    (b) a decrease in current through the junction.
    (c) the current to drop to zero.
    (d) wild fluctuations in the current through the junction.

7. An LED can be useful in communications because
   (a) its intensity remains constant under widely changing input voltages.
   (b) its intensity can be rapidly modulated.
   (c) it does not break down when its avalanche voltage is exceeded.
   (d) it consumes essentially no current.

8. In a conventional diode, once the forward bias across the P-N junction exceeds the forward breakover voltage, a further increase in forward bias causes
   (a) an increase in the current through the junction.
   (b) a decrease in current through the junction.
   (c) the current to drop to zero.
   (d) wild fluctuations in the current through the junction.

9. A thyrector can be used to
   (a) adjust the frequency of a voltage-controlled oscillator.
   (b) generate coherent IR energy.
   (c) receive modulated-light signals.
   (d) suppress transients in a power supply.

10. The junction capacitance of a varactor, when the device is operating under normal conditions, depends on
   (a) the reverse bias voltage.
   (b) the forward breakover voltage.
   (c) the current through the device.
   (d) the avalanche voltage.

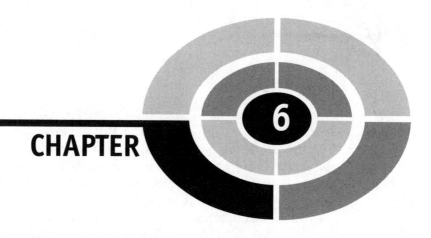

Transistors and Integrated Circuits

Transistors can generate signals, change weak signals into strong ones, mix signals, and act as high-speed switches. *Integrated circuits* (ICs) can perform countless functions. An IC contains many transistors, along with other components, on a single "chip" of semiconductor material.

The Bipolar Transistor

Bipolar transistors have three sections of semiconductor material with two P-N junctions. There are two geometries: a P type layer between two N type layers (the *NPN transistor*), or an N type layer between two P type layers (the *PNP transistor*).

NPN AND PNP

A simplified drawing of an NPN transistor, and its schematic symbol, are shown in Figs. 6-1 A and B. The P type, or center, layer forms the *base*. The thinner of the N type semiconductors is the *emitter*, and the thicker is the *collector*. Sometimes these are labeled B, E, and C in schematic diagrams, although the transistor symbol itself indicates which is which. (The arrow, which points outward, is at the emitter.)

A simplified drawing of a PNP transistor, and its schematic symbol, are shown in Figs. 6-1 C and D. The N type, or center, layer forms the base. The thinner of the P type semiconductors is the emitter, and the thicker is the collector. The arrow, which points inward, is at the emitter.

In most applications, an NPN device can be replaced with a PNP device or vice-versa, and the power-supply polarity reversed, and the circuit will still work if the new device has the appropriate specifications.

There are various kinds of bipolar transistors. Some are used for RF amplifiers and oscillators; others are intended for audio frequencies (AF). Some can handle high power, and others are made for weak-signal work. Some are manufactured for switching, and others are intended for signal processing.

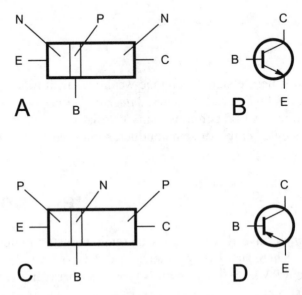

Fig. 6-1. Pictorial diagram of NPN transistor (A), schematic symbol for NPN transistor (B), pictorial diagram of PNP transistor (C), and schematic symbol for PNP transistor (D).

NPN POWER CONNECTIONS

The normal method of applying DC to the emitter and collector of an NPN transistor is to have the emitter negative and the collector positive. This is shown by the connection of the battery in Fig. 6-2. Typical voltages are between about 3 V and 50 V.

The base is labeled "control" because, with constant collector-to-emitter voltage (labeled E_C or V_C), the flow of current through the transistor depends on the current that flows at the emitter-base junction (called the *base current* and labeled I_B). What happens at the base is critical to what happens in the whole transistor.

ZERO BIAS

When the base is not connected to anything, or when it is short-circuited to the emitter for DC, a bipolar transistor is said to be at *zero bias*. No appreciable current can flow between the emitter and the collector unless there is a forward bias at least equal to the *forward breakover* voltage at the emitter-base (E-B) junction. For silicon transistors, this is 0.6 V; for germanium transistors, it is 0.3 V.

With zero bias, the base current I_B is zero, and the E-B junction does not conduct. This prevents current from flowing in the collector circuit unless a signal is injected at the base to change the situation. In an NPN transistor, this signal must have a positive polarity, and its peaks must be sufficient to overcome the forward breakover of the E-B junction for at least a portion of the input signal cycle.

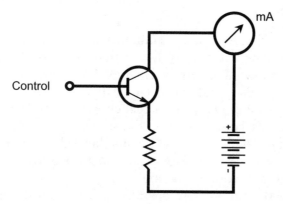

Fig. 6-2.   Power supply connection to the emitter and collector of an NPN transistor.

REVERSE BIAS

Suppose another battery is connected to the base of the NPN transistor at the point marked "control," so the base is negative with respect to the emitter. The addition of this new battery will cause the E-B junction to be in a condition of *reverse bias*. It is assumed that this new battery is not of such a high voltage that avalanche breakdown takes place at the E-B junction.

When the E-B junction of a transistor is reverse-biased, no current flows between the emitter and the collector. A signal might be injected to overcome the reverse-bias battery and the forward-breakover voltage of the E-B junction, but such a signal must have positive voltage peaks high enough to cause conduction of the E-B junction for part of the input signal cycle. Otherwise the device will remain *cut off* for the entire cycle.

FORWARD BIAS

Suppose the bias at the base of an NPN transistor is positive relative to the emitter, starting at small levels and gradually increasing. This is *forward bias*. If this bias is less than forward breakover, no current flows. But once the voltage reaches breakover, the E-B junction begins to conduct.

Despite reverse bias at the base-collector (B-C) junction, the emitter-collector current, more often called *collector current* and denoted I_C, flows when the E-B junction conducts. A small rise in the positive-polarity signal at the base, attended by a small rise in the base current I_B, will cause a large increase in I_C. This is known as *current amplification*, and is the effect that makes it possible for a bipolar transistor to amplify signals.

SATURATION

If I_B continues to rise, a point is eventually reached where I_C increases much less rapidly. Ultimately, the I_C vs. I_B function, or *characteristic curve*, levels off. Figure 6-3 shows a *family of characteristic curves* for a hypothetical bipolar transistor. Each individual curve in the set is characteristic of a certain collector voltage (E_C). The actual current levels depend on the particular device; they are larger for power transistors and smaller for weak-signal transistors. (That's why the axes are not labeled with numbers.) Where the curves level off, the transistor is in a state of *saturation*. Under these conditions it will not work as an amplifier.

Bipolar transistors are rarely operated in the saturated state in analog circuits. They are sometimes biased to saturation in digital circuits where the

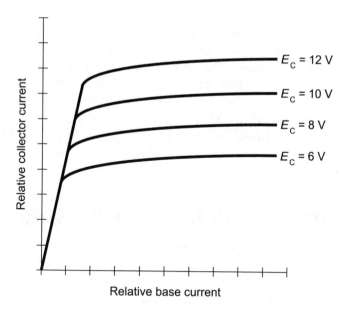

Fig. 6-3. A family of characteristic curves for a bipolar transistor.

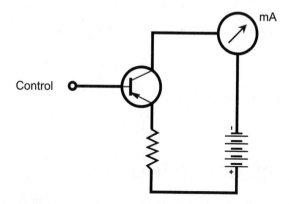

Fig. 6-4. Power supply connection to the emitter and collector of a PNP transistor.

signal state is either "all the way on" (high, or logic 1) or "all the way off" (low, or logic 0).

PNP POWER CONNECTIONS

For a PNP transistor, the DC power supply connection to the emitter and collector is a mirror image of the case for an NPN device, as shown in Fig. 6-4. The polarity of the battery is reversed, compared with the

NPN circuit. To overcome forward breakover at the E-B junction, an applied voltage or signal at the base must attain sufficient negative peak polarity.

Either type of transistor, the PNP or the NPN, can serve as a *current valve*. Small changes in the base current, I_B, induce large fluctuations in the collector current, I_C, when the device is operated in the region of the characteristic curve where the slope is steep. The internal atomic activity is a little different in the PNP device as compared with the NPN device, but the performance of the external circuitry is the same in most practical situations.

STATIC FORWARD CURRENT TRANSFER RATIO

The largest possible current amplification factor for a particular transistor is known as its *beta* (symbolized β), and can range from a few times up to hundreds of times. One way to express the beta of a transistor is as the *static forward current transfer ratio*, abbreviated H_{FE}. This is the ratio of collector current to base current. Mathematically,

$$H_{FE} = I_C/I_B$$

For example, if a base current, I_B, of 1 mA results in a collector current, I_C, of 35 mA, then:

$$H_{FE} = 35/1 = 35$$

If $I_B = 0.5$ mA and $I_C = 35$ mA, then:

$$H_{FE} = 35/0.5 = 70$$

DYNAMIC CURRENT AMPLIFICATION

Another way to express current amplification is as the ratio of a change in I_C to a small change in I_B that produces it. This is the *dynamic current amplification*, also known as *current gain*. Let's symbolize the words "the change in" by the uppercase Greek letter Δ (delta). Then, according to this second definition,

$$\text{Current gain} = \Delta I_C/\Delta I_B$$

The ratio $\Delta I_C/\Delta I_B$ is greatest where the slope of the characteristic curve is steepest.

When a transistor is operated at a point represented by the steepest part of its characteristic curve, the device provides the largest possible current gain, as long as the input signal is small. This value is close to H_{FE}. Because the characteristic curve is almost a straight line in this region, the transistor can serve as a *linear amplifier* if the input signal is not too strong. In a linear amplifier, the output signal wave looks the same as the input signal wave, except that the output signal is stronger.

If the operating point is not in the straight-line portion of the characteristic curve, the current gain decreases and the amplifier becomes *nonlinear*. The same thing can happen if the input signal is strong enough to drive the transistor out of the straight-line part of the curve during any portion of the signal cycle. In a *nonlinear circuit*, the output signal wave is shaped differently than the input signal wave. The result is distortion of the signal waveform. In some situations this can be tolerated; in other cases it means trouble.

GAIN VERSUS FREQUENCY

In a bipolar transistor, the current gain decreases as the signal frequency increases. There are two expressions for current-gain-versus-frequency behavior.

The *gain bandwidth product*, abbreviated f_T, is the frequency at which the current gain becomes equal to unity (1) with the emitter connected to ground. The *alpha cutoff* is the frequency at which the gain becomes 0.707 times its value at 1 kHz. A transistor can have current gain at frequencies above its alpha cutoff, but it cannot produce gain at frequencies higher than f_T.

PROBLEM 6-1
What is the static forward current transfer ratio in a device where a base current of 3 mA produces a collector current of 25 mA?

SOLUTION 6-1
Use the formula for the static forward current transfer ratio, H_{FE}:

$$H_{FE} = I_C / I_B$$

In this case, the base current, I_B, is 3 mA and the collector current, I_C, is 25 mA. Therefore:

$$H_{FE} = 25/3 = 8.33$$

PROBLEM 6-2
What is the current gain in a transistor where a change in the base current of 0.1 μA (0.1 microampere or 0.0000001 ampere) results in a change in the collector current of 1 μA?

SOLUTION 6-2

Remember the formula for current gain:

$$\text{Current gain} = \Delta I_C / \Delta I_B$$

In this case, $\Delta I_C = 1\,\mu A$ and $\Delta I_B = 0.1\,\mu A$. Therefore:

$$\text{Current gain} = 1/0.1 = 10$$

Basic Bipolar Transistor Circuits

A bipolar transistor can be connected in a circuit in three general ways. The emitter can be grounded with respect to the signal, the base can be grounded with respect to the signal, or the collector can be grounded with respect to the signal. These are called *common-emitter*, *common-base*, and *common-collector* circuits, respectively.

COMMON EMITTER

The basic common-emitter circuit configuration is shown in Fig. 6-5. Capacitor C_1 presents a short circuit to the AC signal, so the emitter is at

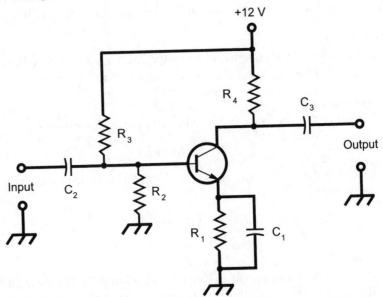

Fig. 6-5.   A common-emitter circuit.

signal ground. Resistor R_1 causes the emitter to have a certain positive DC voltage with respect to ground (or a negative voltage, if a PNP transistor is used). The exact DC voltage at the emitter depends on the value of R_1, and on the bias voltage, which is determined by the ratio of the values of resistors R_2 and R_3. This bias can be anything from zero, or ground potential, up to the supply voltage. Normally, it is a couple of volts.

Capacitors C_2 and C_3 block DC to or from the input and output, while letting the AC signal pass. Resistor R_4 keeps the output signal from being shorted out through the power supply.

A signal voltage enters the common-emitter circuit through C_2, where it causes the base current, I_B, to vary. Small fluctuations in I_B cause large variations in the collector current, I_C. This current passes through R_4, causing a fluctuating DC voltage to appear across this resistor. The AC part of this passes through C_3 to the output.

The common-emitter configuration is capable of producing the greatest gain of any arrangement. The AC output wave is 180° out of phase with the input wave. But if circuitry outside the transistor reverses the phase again (and this is often the case), this type of circuit can generate a signal of its own. Then it is said to *break into oscillation*. Under these conditions, the circuit fails as an amplifier, and can cause all sorts of bad things to happen in an electronic system. Oscillation in a common-emitter amplifier can be prevented by making sure that the gain is not too great. Alternatively, another circuit configuration can be used.

COMMON BASE

The common-base circuit (Fig. 6-6) places the base at signal ground. The DC bias on the transistor is the same for this circuit as for the common-emitter circuit. The input signal is applied at the emitter. This causes fluctuations in the voltage across R_1, causing variations in I_B. The result of these small current fluctuations is a large change in the current through R_4. Therefore, amplification occurs. In the common-base arrangement, the output signal is in phase with the input.

The signal enters through C_1. Resistor R_1 keeps the input signal from being shorted to ground. Bias is provided by R_2 and R_3. Capacitor C_2 keeps the base at signal ground. Resistor R_4 keeps the signal from being shorted out through the power supply. The output signal goes through C_3.

A common-base circuit can't produce as much gain as a common-emitter circuit, but is less prone to oscillate. It is less susceptible to the undesirable effects of feedback, which can cause an amplifier to "rage out of control" in

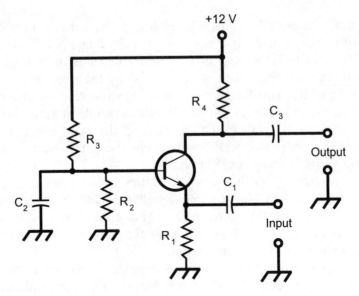

Fig. 6-6. A common-base circuit.

much the same way as a public-address system "howls" when the microphone picks up too much sound from the speakers. For this reason, common-base circuits are popular as high-power amplifiers such as those used in radio transmitters.

COMMON COLLECTOR

A common-collector circuit (Fig. 6-7) operates with the collector at signal ground. The input is applied at the base.

The signal passes through C_2 onto the base of the transistor. Resistors R_2 and R_3 provide the bias. Resistor R_4 limits the current through the transistor. Capacitor C_3 keeps the collector at signal ground. A fluctuating current flows through R_1, and a fluctuating voltage therefore appears across it. The AC component passes through C_1 to the output. Because the output follows the emitter current, this circuit is sometimes called an *emitter follower*.

The output of this circuit is in phase with the input. When well designed, an emitter follower works over a wide range of frequencies, and is a low-cost alternative to an RF signal transformer. This type of circuit is not used as an amplifier, but it can help to provide isolation between two different parts of an electronic system. This sort of isolation circuit is called a *buffer*.

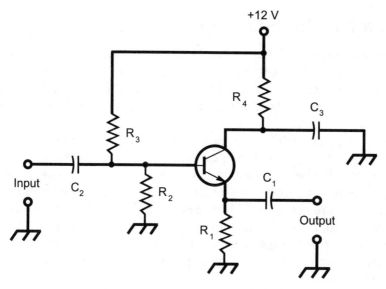

Fig. 6-7.   A common-collector circuit.

The Field-Effect Transistor

The other major category of transistor, besides the bipolar device, is the *field-effect transistor (FET)*. There are two main types of FET: the *junction FET (JFET)* and the *metal-oxide-semiconductor FET (MOSFET)*.

PRINCIPLE OF THE JFET

In a JFET, the current varies because of the effects of an electric field within the device. Electrons or holes (electron deficiencies, or vacant spaces in atoms where electrons normally belong) flow along a *channel* from the *source* (S) electrode to the *drain* (D) electrode. This results in a drain current, I_D, that is normally the same as the source current, I_S. The current through the channel depends on the voltage at the *gate* (G).

Small changes in the gate voltage, E_G, can cause large changes in the current through the channel, and this ultimately appears as large changes in the drain current, I_D. These current fluctuations, when passing through an external resistance, cause large fluctuations in the voltage across that resistance. This is how an FET can produce *voltage amplification*.

N-CHANNEL AND P-CHANNEL

A simplified drawing of an *N-channel JFET*, and its schematic symbol, are shown in Figs. 6-8 A and B. The N type material forms the path for the current. Most of the charge carriers are electrons; engineers would say these are the *majority carriers*. The drain is connected to the positive power-supply terminal.

In an N-channel JFET, the gate consists of P type material. Another, larger section of P type material, called the *substrate*, forms a boundary on the side of the channel opposite the gate. The voltage on the gate produces an electric field that interferes with the flow of charge carriers through the channel. The more negative E_G becomes, the more the electric field chokes off the current though the channel, and the smaller I_D becomes.

A *P-channel JFET* (Figs. 6-8 C and D) has a current pathway made of P type semiconductor material. The majority carriers are holes. The drain is connected to the negative power-supply terminal. The gate and substrate are made of N type material. The more positive E_G gets, the more the electric field chokes off the current through the channel, and the smaller I_D becomes.

An N-channel JFET can be recognized in schematic diagrams by an arrow pointing inward at the gate, and a P-channel JFET can be recognized by an

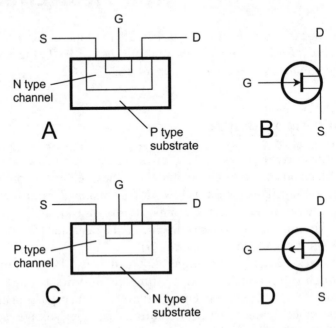

Fig. 6-8.   Pictorial diagram of N-channel JFET (A), symbol for N-channel JFET (B), pictorial diagram of P-channel JFET (C), and symbol for P-channel JFET (D).

arrow pointing outward at the gate. In some diagrams these arrows are missing, but the power-supply polarity gives away which type of device is used. When the drain goes to the positive power-supply voltage (with the negative power-supply terminal usually connected to ground), it indicates an N-channel JFET. When the drain goes to the negative power-supply voltage (with the positive power-supply terminal usually connected to ground), it indicates a P-channel JFET.

An N-channel JFET can almost always be replaced with a P-channel JFET, and the power-supply polarity reversed, and the circuit will still work if the new device has the right specifications.

DEPLETION AND PINCHOFF

A JFET works because the voltage at the gate causes an electric field that interferes, more or less, with the flow of charge carriers along the channel.

As the drain voltage E_D increases, so does the drain current I_D, up to a certain leveling-off value. This is true as long as the gate voltage E_G is constant, and is not too large negatively. As E_G increases (negatively in an N channel or positively in a P channel), a *depletion region* begins to form in the channel. Charge carriers cannot flow in the depletion region, so they are forced to pass through a narrowed channel.

The larger E_G becomes, the wider the depletion region gets, and the narrower the channel becomes. If E_G gets high enough, the depletion region gets so large that it closes off the channel altogether. This prevents any flow of charge carriers from the source to the drain. When this occurs, the condition is called *pinchoff*.

JFET BIASING

Two biasing arrangements for an N-channel JFET are shown in Fig. 6-9. At A, the gate is grounded through resistor R_2. The source resistor, R_1, limits the current through the JFET. Resistors R_1 and R_2 determine the gate bias. The drain current, I_D, flows through R_3, producing a voltage across this resistor. The AC output signal passes through C_2.

At B, the gate is connected through potentiometer R_2 to a negative supply voltage. Adjusting this potentiometer produces variable negative gate voltage E_G between R_2 and R_3. Resistor R_1 limits the current through the JFET. The drain current, I_D, flows through R_4, producing a voltage across it. The AC output signal passes through C_2.

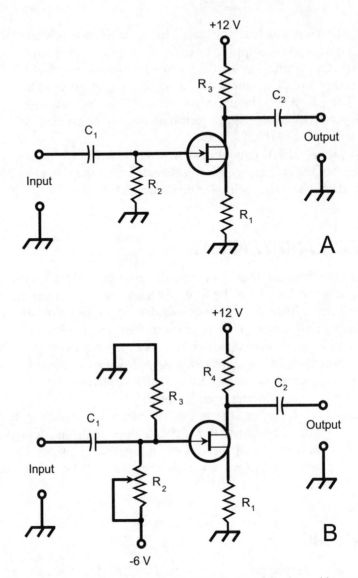

Fig. 6-9. Two methods of biasing an N-channel JFET. At A, fixed gate bias; at B, variable gate bias.

In both of these circuits, the drain is positive relative to ground. In the case of a P-channel JFET circuit, the polarities must be reversed.

The biasing arrangement in Fig. 6-9A is commonly used for weak-signal amplifiers, low-level amplifiers, and oscillators. The scheme at B is employed in certain power amplifiers having a substantial input signal. Typical JFET power-supply voltages are comparable to those with bipolar transistors. The voltage between the source and drain, E_D, can range from approximately 3 V to 50 V; most often it is 6 V to 12 V.

HOW THE JFET AMPLIFIES

Figure 6-10 shows the relative drain (channel) current, I_D, as a function of the gate bias voltage, E_G, for a hypothetical N-channel JFET. The drain voltage, E_D, is assumed to be constant.

When E_G is fairly large and negative, the JFET is pinched off, and no current flows through the channel. As E_G gets less negative, the channel opens up, and current begins flowing. As E_G gets still less negative, the channel gets wider, and the current I_D through the device increases. As E_G approaches the point where the source-gate (S-G) junction is at forward breakover, the channel conducts as well as it possibly can; it is "wide open." If E_G gets more positive still, reaching the forward breakover voltage and causing the S-G junction to conduct, some of the current in the channel leaks out through the gate. This is usually an undesirable condition.

The greatest amplification for weak signals occurs when E_G is such that the slope of the curve in Fig. 6-10 is steepest. This is shown by the range marked X in the graph. For high-power amplification when the input signal is already quite strong, results are often the best when the JFET is biased at or beyond pinchoff, in the range marked Y.

In a practical JFET amplifier circuit, the drain current passes through the drain resistor, as shown in Fig. 6-9 A or B. Small fluctuations in E_G cause

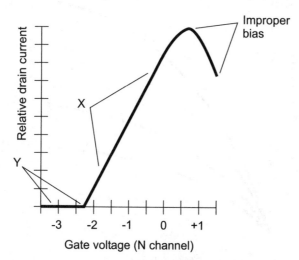

Fig. 6-10.   Relative drain current as a function of gate voltage for a hypothetical N-channel JFET.

large changes in I_D, and these variations in turn produce wide swings in the DC voltage across R_3 (at A) or R_4 (at B). The AC part of this voltage goes through capacitor C_2, and appears at the output as a signal of much greater peak voltage than that of the input signal at the gate.

DRAIN CURRENT VERSUS DRAIN VOLTAGE

In any JFET, the drain current I_D can be plotted as a function of drain voltage E_D for various values of gate voltage E_G. The resulting graph is called a *family of characteristic curves* for the device. Figure 6-11 shows a family of characteristic curves for a hypothetical N-channel JFET. Also of importance is the curve of I_D vs. E_G (one example of which is shown in Fig. 6-10).

TRANSCONDUCTANCE

The JFET expression for current gain is called *dynamic mutual conductance* or *transconductance*.

Look at Fig. 6-10, and suppose that the gate voltage E_G is a certain value, with a drain current I_D resulting. If the gate voltage changes by a small amount ΔE_G, then the drain current also changes by a certain increment ΔI_D. The transconductance is the ratio $\Delta I_D/\Delta E_G$, the slope of the curve in that region.

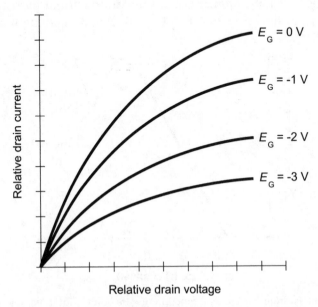

Fig. 6-11. A family of characteristic curves for a hypothetical N-channel JFET.

The value of $\Delta I_D / \Delta E_G$ is not the same everywhere along the curve. When the JFET is biased beyond pinchoff, in the region marked Y in the graph, the curve is horizontal. There is no fluctuation in drain current, even if the gate voltage changes. Only when the channel is conducting, is there a change in I_D when there is a change in E_G. The region where the transconductance is the greatest is the region marked X, where the slope is steepest. This is where the most gain (amplification) can be obtained.

PROBLEM 6-3
Refer to Fig. 6-11. Suppose the device described by this family of curves is operated with a DC gate voltage $E_G = -1$ V. Suppose we define and measure the "effective resistance" R of this device by dividing the DC drain voltage by the drain current. What happens to R as the drain voltage E_D increases?

SOLUTION 6-3
Mathematically, the value of R is given by this equation:

$$R = E_D / I_D$$

This ratio is smallest at small drain voltages, and gradually increases as E_D increases. This is a result of the fact that the drain current (or current through the device) increases less and less rapidly as the drain voltage increases.

PROBLEM 6-4
Look again at Fig. 6-10. What happens to the transconductance as the gate voltage becomes less and less negative, passing through 0 and then increasing positively?

SOLUTION 6-4
The transconductance at any gate voltage E_G is equal to the slope of the curve at the point corresponding to that E_G. At first this is large and positive because the curve ramps upward. This is the region marked X in the graph. When the gate voltage reaches a certain point, the transconductance starts to decrease, and reaches 0 when E_G is a little less than $+1$ V. Beyond this point, the transconductance actually becomes negative, because a further positive increase in E_G causes the drain current I_D to decline.

The MOSFET

The acronym MOSFET (pronounced "MOSS-fet") stands for *metal-oxide-semiconductor field-effect transistor*. This type of component can be

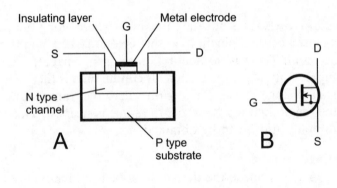

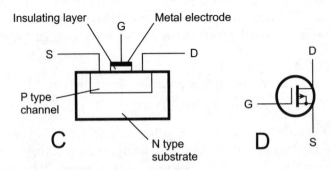

Fig. 6-12. Pictorial diagram of N-channel MOSFET (A), schematic symbol for N-channel MOSFET (B), pictorial diagram of P-channel MOSFET (C), and schematic symbol for P-channel MOSFET (D).

constructed with a channel of N type material, or with a channel of P type material. The former type is called an *N-channel MOSFET*; the latter type is called a *P-channel MOSFET*.

A simplified cross-sectional drawing of an N-channel MOSFET, along with the schematic symbol, is shown in Figs. 6-12 A and B. The P-channel cross-sectional drawing and symbol are shown at C and D.

THE INSULATED GATE

When the MOSFET was first developed, it was called an *insulated-gate FET* or IGFET. This is perhaps more descriptive of the device than the currently accepted name. The gate electrode is actually insulated, by a thin layer of dielectric material, from the channel.

The input impedance for a MOSFET is even higher than that of a JFET when the input is applied at the gate electrode. In fact, the gate-to-source resistance of a typical MOSFET is comparable to that of a well-designed capacitor; it is practically infinite.

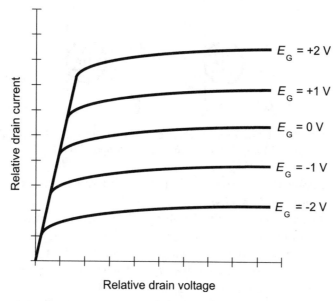

Fig. 6-13. A family of characteristic curves for a hypothetical N-channel MOSFET.

A family of characteristic curves for a hypothetical N-channel MOSFET is shown in Fig. 6-13. Note that the curves rise steeply at first, but then, as the positive drain voltage increases beyond a certain threshold, the curves level off much more quickly than they do for the JFET.

ELECTROSTATIC DISCHARGE

One of the most serious shortcomings of MOSFETs is the fact that they are easily destroyed by electrostatic discharge. When circuits containing metal-oxide-semiconductor devices are built or serviced, technicians must use special equipment to ensure that their hands do not carry electrostatic charges that might ruin the components. If a static discharge occurs through the thin, fragile dielectric layer inside a MOSFET, the damage is permanent.

DEPLETION VERSUS ENHANCEMENT

In a JFET, the channel conducts with zero gate bias, that is, when the gate is at the same potential as the source ($E_G = 0$). As E_G increases negatively, the depletion region grows, and the charge carriers are forced pass through a narrowed channel. This is known as *depletion mode*. A MOSFET can also be made to work in the depletion mode. The drawings and schematic symbols

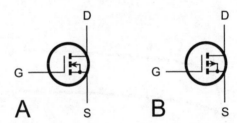

Fig. 6-14.   At A, the symbol for an N-channel enhancement-mode MOSFET. At B, the symbol for a P-channel enhancement-mode MOSFET.

of Fig. 6-12 depict the internal construction and schematic symbols for the so-called *depletion-mode MOSFET*.

Metal-oxide-semiconductor technology allows a different mode of operation, as well. An *enhancement-mode MOSFET* has a pinched-off channel at zero bias. It is necessary to apply a gate bias voltage, E_G, to create a channel in this type of device. If $E_G = 0$, then $I_D = 0$ when there is no signal input. This is because of differences in the internal construction of enhancement-mode MOSFETs, compared with depletion-mode MOSFETs.

The schematic symbols for N-channel and P-channel enhancement-mode MOSFETs are shown in Fig. 6-14. The right-hand vertical lines inside the symbols are broken, rather than solid as is the case in the symbol for the depletion-mode MOSFET. This is how you can distinguish between the two types of devices when you see them in circuit diagrams.

Basic FET Circuits

There are three general circuit configurations for FETs, whether they be of the junction (J) variety or the metal-oxide-semiconductor (MOS) variety. These three arrangements have the source, the gate or the drain at signal ground, respectively. They are the FET equivalents of the common-emitter, common-base, and common-collector bipolar-transistor circuits.

COMMON SOURCE

In a *common-source circuit*, the input signal is applied to the gate (Fig. 6-15). An N-channel JFET is shown here, but the device could be an N-channel, depletion-mode MOSFET and the circuit diagram would be the same. (For an N-channel, enhancement-mode MOSFET, an extra resistor is necessary, connected between the gate and the positive power supply terminal.)

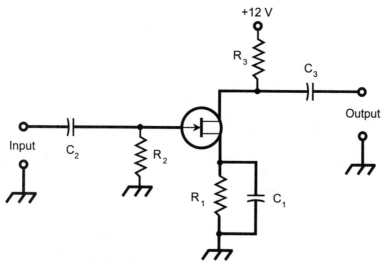

Fig. 6-15. Common-source circuit configuration. Illustration for Problem 6-5.

For P-channel devices, the power supply must provide a negative, rather than a positive, voltage; otherwise the circuit details are the same as those shown in Fig. 6-15.

Capacitor C_1 and resistor R_1 place the source at signal ground, while elevating the source above ground for DC. The AC signal enters through C_2; resistor R_2 adjusts the input impedance and provides bias for the gate. The AC signal passes out of the circuit through C_3. Resistor R_3 keeps the output signal from being shorted out through the power supply.

The circuit of Fig. 6-15 is the basis for amplifiers and oscillators, especially at RF. The common-source arrangement provides the greatest gain of the three FET circuit configurations. The output is 180° out of phase with the input.

COMMON GATE

The *common-gate circuit* (Fig. 6-16) has the gate at signal ground. The input is applied to the source. The illustration shows an N-channel JFET. For other types of FETs, the same considerations apply as described above for the common-source circuit. Enhancement-mode devices require a resistor between the gate and the positive supply terminal (or the negative terminal if the MOSFET is P-channel).

The DC bias for the common-gate circuit is basically the same as that for the common-source arrangement, but the signal follows a different path. The

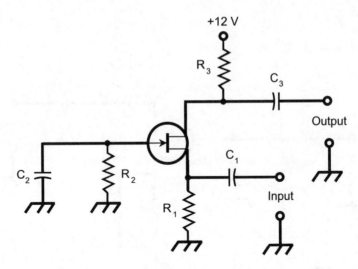

Fig. 6-16. Common-gate circuit configuration. Illustration for Problem 6-6.

AC input signal enters through C_1. Resistor R_1 keeps the input from being shorted to ground. Gate bias is provided by R_1 and R_2. Capacitor C_2 places the gate at signal ground. In some common-gate circuits, the gate electrode is directly grounded, and components R_2 and C_2 are not used. The output signal exits the circuit through C_3. Resistor R_3 keeps the output signal from being shorted through the power supply.

The common-gate arrangement produces less gain than its common-source counterpart. But a common-gate amplifier is not likely to break into unwanted oscillation. The output is in phase with the input.

COMMON DRAIN

A *common-drain circuit* is shown in Fig. 6-17. This circuit has the collector at signal ground. It is sometimes called a *source follower*.

The FET is biased in the same way as for the common-source and common-gate circuits. In the illustration, an N-channel JFET is shown, but any other kind of FET can be used, reversing the polarity for P-channel devices. Enhancement-mode MOSFETs require a resistor between the gate and the positive supply terminal (or the negative terminal if the MOSFET is P-channel).

The input signal passes through C_2 to the gate. Resistors R_1 and R_2 provide gate bias. Resistor R_3 limits the current. Capacitor C_3 keeps the drain at signal ground. Fluctuating DC (the channel current) flows through

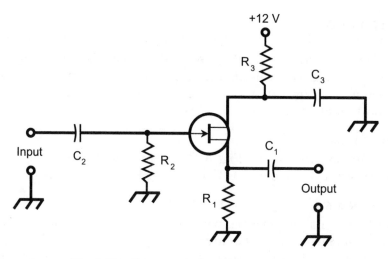

Fig. 6-17.   Common-drain circuit configuration.

R_1 as a result of the input signal; this current causes a fluctuating DC voltage to appear across the resistor. The output is taken from the source, and its AC component passes through C_1.

The output of the common-drain circuit is in phase with the input. This circuit, like the emitter follower, is often used as a low-cost alternative to a transformer, especially in RF applications.

INTERCHANGEABILITY

In the preceding discussions, accompanied by Figs. 6-15, 6-16, and 6-17, you might get the impression that JFETs and MOSFETs are directly interchangeable. This is sometimes true, but not always.

Although the circuit diagrams for JFET and depletion-mode MOSFET devices might look identical when component values aren't specified, there's often a difference in the optimum resistances and capacitances. These can vary not only between JFETs and MOSFETs, but also between depletion-mode MOSFETs and enhancement-mode MOSFETs. It is beyond the scope of this course to get into specific design examples. But you should remember that the two types of devices are sometimes, but not always, directly interchangeable.

PROBLEM 6-5
In the common-source circuit (Fig. 6-15), what would happen to the output signal if resistor R_2 were to short out? What would happen to the gate bias?

SOLUTION 6-5

Resistor R_2 is connected in parallel with the input signal. This resistor also provides bias for the gate. If R_2 were to short out, the input signal would disappear from the gate, causing the output signal to drop to zero. In addition, it would alter the gate bias, so the FET might no longer operate in the proper portion of its characteristic curve.

PROBLEM 6-6

In the common-gate circuit (Fig. 6-16), what would happen to the drain current if resistor R_1 were to open? What would happen to the output signal?

SOLUTION 6-6

Resistor R_1 is connected in series with the current path through the channel. Thus, if R_1 were to open up, the current through the device (and therefore the drain current) would be interrupted. This would prevent the FET from amplifying the signal. A little of the input signal might leak through to the output, but even if this did occur, the output signal would be weak.

Integrated Circuits

Most *integrated circuits* (ICs) look like plastic boxes with protruding metal pins. The schematic symbol for an IC is a triangle or rectangle with the component designator written inside.

A *linear IC* is used to process analog signals such as voices, music, and most RF transmissions. The term "linear" arises from the fact that the instantaneous output is a straight-line mathematical function of the instantaneous input. *Digital ICs* consist of gates that perform logical operations at high speeds. These devices are definitely not linear!

COMPACTNESS

Integrated-circuit devices and systems are far more compact than equivalent circuits made from *discrete components* – that is, individual resistors, capacitors, and transistors. More complex circuits can be built, and kept down to a reasonable size, using ICs as compared with discrete components. That is why, for example, there are notebook computers available today that have capabilities more advanced than the most powerful "supercomputers" built in the middle of the last century.

HIGH SPEED

In an IC, the interconnections among components are physically tiny, making high switching speeds possible. Electric currents travel fast, but not instantaneously. The faster the charge carriers move from one component to another, the more operations can be performed per unit time, and the less time is required for complex operations.

LOW POWER REQUIREMENT

Integrated circuits generally use less power than equivalent discrete-component circuits. This is important if batteries are used. Because ICs draw less current, they produce less heat than their discrete-component equivalents. This results in better efficiency, and minimizes problems that plague equipment that gets hot with use, such as frequency drift and generation of internal noise.

Despite this advantage, certain types of ICs generate enough heat, in proportion to their small physical size, to require cooling fans or other means to keep them from getting too hot. A good example is the microprocessor in a personal computer.

RELIABILITY

Systems using ICs fail less often, per component-hour of use, than systems that make use of discrete components. This is mainly because all interconnections are sealed within an IC case, preventing corrosion or the intrusion of dust. The reduced failure rate translates into less downtime.

EASE OF MAINTENANCE

Integrated-circuit technology lowers service costs, because repair procedures are simple when failures occur. Many systems use sockets for ICs, and replacement is simply a matter of finding the faulty IC, unplugging it, and plugging in a new one. Special desoldering equipment is used for servicing circuit boards that have ICs soldered directly to the foil.

MODULAR CONSTRUCTION

Modern IC appliances employ *modular construction*. Individual ICs perform defined functions within a circuit board. The circuit board or card, in turn, fits into a socket and has a specific purpose. Computers, programmed with customized software, are used by technicians to locate the faulty card in a system. The card can be pulled and replaced, getting the system back to the user in the shortest possible time.

INDUCTORS IMPRACTICAL

Devices using ICs must generally be designed to work without inductors, because inductances cannot easily be fabricated onto silicon chips. Resistance-capacitance (*RC*) circuits are capable of doing most things that inductance-capacitance (*LC*) circuits can do. Therefore, some inductances can be replaced with resistances, which are easily fabricated onto silicon chips.

MEGA-POWER IMPOSSIBLE

High-power amplifiers cannot, in general, be built onto semiconductor chips. High power necessitates a certain minimum physical bulk and mass to allow the conduction and radiation of excess heat energy. Power transistors and, in some systems, vacuum tubes are generally employed for high-power amplification.

OPERATIONAL AMPLIFIER

An *operational amplifier* ("op amp") is a linear IC that consists of several transistors, resistors, diodes, and capacitors, interconnected to produce high gain over a wide range of signal input frequencies. An op amp has two inputs and one output.

When a signal is applied to the *noninverting input*, the output is in phase with the input. When a signal is applied to the *inverting input*, the output is $180°$ out of phase with the input. An op amp has two power supply connections, one for the emitters of the transistors (V_{ee}) and one for the collectors (V_{cc}). The symbol for an op amp is a triangle. The inputs, output and power supply connections are drawn as lines emerging from the triangle.

The gain of an op amp is determined by external resistors. Normally, a resistor is connected between the output and the inverting input. This is the *closed-loop configuration*. The feedback is negative or out of phase, causing the gain to be less than it would be if there were no feedback (*open-loop*

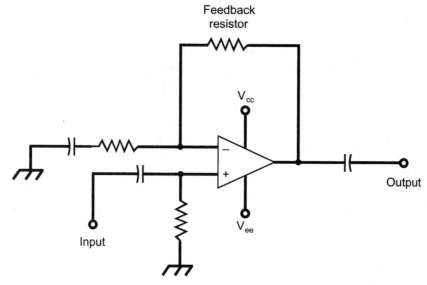

Fig. 6-18.   Closed-loop op amp circuit.

configuration). A basic closed-loop amplifier using an op amp is shown in Fig. 6-18. If the resistor is connected between the output and the noninverting input, positive feedback occurs. If the resistor is set for the correct value, oscillation takes place.

When a resistance-capacitance (*RC*) combination is used in the negative feedback loop of an op amp, the amplification factor varies with the frequency. It is possible to get a *lowpass response* (gain decreases as the frequency increases), a *highpass response* (gain increases as the frequency increases), a *resonant peak* (gain reaches a maximum at a specific frequency), or a *resonant notch* (gain reaches a minimum at a specific frequency) using an op amp and various *RC* feedback arrangements.

If an *RC* combination is used in a positive-feedback loop with the intention of getting an oscillator, the resistance and capacitance values determine the frequency at which the circuit oscillates.

VOLTAGE REGULATOR

A *voltage regulator* is a linear IC that acts to control the output voltage of a power supply. This is important with precision electronic equipment. These ICs are available with various voltage and current ratings. Typical voltage regulator ICs have three terminals. Because of this, they are sometimes mistaken for large transistors by casual observers.

TIMER

A *timer* IC is a form of oscillator. It produces a delayed output, with the delay being variable to suit the needs of a particular device. The delay is generated by counting the number of oscillator pulses. The length of the delay is adjusted by means of external resistors and capacitors.

MULTIPLEXER AND DEMULTIPLEXER

A *multiplexer* is a linear IC that allows several different signals to be combined in a single communications channel. A *demultiplexer* separates the individual signals from a channel containing multiple signals. A *multiplexer/demultiplexer* is an IC that can perform either of these functions.

COMPARATOR

A *comparator* IC has two inputs. The device compares the voltages at the two inputs (called A and B). If the input at A is significantly greater than the input at B, the output is about $+5\,$V. This is logic 1, or high. If the input at A is less than or equal to the input at B, the output voltage is about $+2\,$V. This is designated as logic 0, or low.

Comparators are employed to actuate, or *trigger*, other devices such as relays and electronic switching circuits. Some can switch between low and high states at high speed; others are slow. Some have low input impedance, and others have high impedance. Some are intended for audio or low-frequency use; others are fabricated for video or high-frequency applications.

RANDOM-ACCESS MEMORY (RAM)

A *random-access memory* (RAM) chip is a digital IC that stores binary data in arrays. The data can be addressed from anywhere in the matrix. Data is easily changed and stored back in RAM.

In most RAM chips, data vanishes on powering down unless some provision is made for *memory backup*. Any memory medium whose content disappears when power is removed is called *volatile memory*. If data is retained when power is removed, then it is called *nonvolatile memory*.

READ-ONLY MEMORY (ROM)

Read-only memory (ROM) can be accessed in whole or in part, but not written over in the course of normal operation. A standard ROM chip, which is a digital device, is programmed at the factory. This permanent programming is known as *firmware*. There are also ROMs that you can program and reprogram yourself.

An *erasable programmable ROM* (EPROM) is a digital IC whose memory is of the read-only type, but that can be reprogrammed by exposure to ultraviolet (UV) radiation. The IC must be taken from the circuit in which it is used, exposed to the UV for several minutes, and then reprogrammed via a special process. Some EPROM chips are susceptible to erasure by X rays. This fact might explain why some people experience malfunctions in notebook computers and other digital devices after they have passed through airport security scanners.

There are EPROMs that can be erased by electrical means. Such an IC is called an EEPROM, for *electrically erasable programmable ROM*. This type of IC does not have to be removed from then circuit for reprogramming.

Quiz

Refer to the text in this chapter if necessary. Answers are in the back of the book.

1. Electrostatic discharge presents the greatest risk of component damage in circuits that make use of
    (a) bipolar devices.
    (b) NPN devices.
    (c) reverse-biased devices.
    (d) MOS devices.

2. The input signal is applied to the gate electrode in a
    (a) common-emitter circuit.
    (b) common-collector circuit.
    (c) common-gate circuit.
    (d) None of the above.

3. The term *common drain* implies that
    (a) the source is at signal ground.
    (b) the input signal must be applied to the collector.
    (c) the drain is at signal ground.
    (d) the input signal must be applied to the emitter.

4. Positive feedback in an op amp can be accomplished by
    (a) connecting the inverting and noninverting inputs both directly to ground.
    (b) connecting the output to the noninverting input through a resistor.
    (c) connecting the output to the inverting input through a resistor.
    (d) connecting the output directly to ground.

5. The contents of a RAM chip
    (a) cannot be overwritten.
    (b) can be overwritten using UV radiation.
    (c) are usually volatile unless backed up in some way.
    (d) are nonvolatile.

6. An NPN transistor can be used as
    (a) an audio amplifier.
    (b) an RF amplifier.
    (c) a switch.
    (d) Any of the above.

7. In a typical NPN transistor circuit, the power supply is connected so that
    (a) the collector is positive with respect to the emitter.
    (b) the collector is negative with respect to the emitter.
    (c) the collector is at the same potential as the emitter.
    (d) the positive terminal runs directly to the base.

8. When a PNP bipolar transistor is reverse-biased,
    (a) a large current flows in the collector circuit under no-signal conditions.
    (b) a small current flows in the collector circuit under no-signal conditions.
    (c) no current flows in the collector circuit under no-signal conditions.
    (d) Any of the above can be true.

9. When two or more signals are combined in a single communications channel, they can be separated out using
    (a) a junction MOSFET.
    (b) an enhancement-mode MOSFET.
    (c) a timer.
    (d) a multiplexer/demultiplexer.

10. Another name for the H_{FE} of a bipolar transistor is the
    (a) alpha.
    (b) alpha cutoff.
    (c) beta.
    (d) beta cutoff.

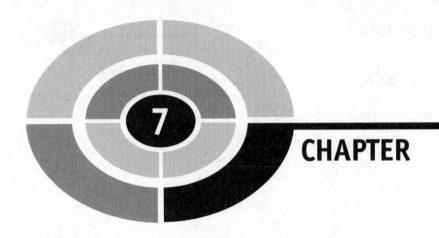

Signal Amplifiers

A *signal amplifier* is a circuit or system that increases the current, voltage, or power level of an audio-frequency (AF) or radio-frequency (RF) signal. In this chapter, we'll examine the most common types of signal amplifiers, and learn how they work.

Amplification Factor

The extent to which a circuit amplifies is called the *amplification factor*. It's usually specified in *decibels* (dB). In any particular circuit, the amplification factor can, and often does, differ considerably as measured for voltage, current, and power.

THE DECIBEL

Humans perceive most variable quantities in a logarithmic manner. Thus, scientists have devised the *decibel system*, in which amplitude variations are expressed according to the *logarithm* of the actual signal strength. *Gain* is assigned positive decibel values; *loss* is assigned negative decibel values.

An amplitude change of 1 dB is roughly equal to the smallest change a listener or observer can detect if the change is expected. If the change is not expected, then the smallest difference a listener or observer can notice is about 3 dB.

VOLTAGE DECIBELS

Imagine an amplifier with an rms AC input voltage of E_{in} and an rms AC output voltage of E_{out}, both specified in the same units. Also assume that the impedances are the same at the input and the output. Then the *voltage gain* of the circuit, in decibels, is given by the formula

$$\text{Gain (dB)} = 20 \log_{10}(E_{out}/E_{in})$$

CURRENT DECIBELS

The *current gain* of a circuit is calculated the same way as is voltage gain. If I_{in} is the rms AC input current and I_{out} is the rms AC output current specified in the same units, and if the input impedance is the same as the output impedance, then

$$\text{Gain (dB)} = 20 \log_{10}(I_{out}/I_{in})$$

Often, a circuit that produces voltage gain will produce current loss, and vice-versa. An example is an AC transformer.

Some circuits have gain for both the voltage and the current, although not the same decibel figures. The reason for the difference is that the output impedance is higher or lower than the input impedance, altering the ratio of voltage to current.

POWER DECIBELS

The *power gain* of a circuit, in decibels, is calculated according to the formula

$$\text{Gain (dB)} = 10 \log_{10}(P_{out}/P_{in})$$

where P_{out} is the output signal power and P_{in} is the input signal power, specified in the same units. This formula holds whether or not the input and output impedances are the same.

PROBLEM 7-1

If the rms AC input voltage to a circuit is 10 millivolts (mV) and the rms AC output voltage is 70 mV, what is the voltage gain? Assume the input and output impedances are identical.

SOLUTION 7-1

Here, we can assign $E_{in} = 10$ and $E_{out} = 70$. Use the above formula for voltage gain:

$$\begin{aligned}
\text{Gain (dB)} &= 20 \log_{10}(E_{out}/E_{in}) \\
&= 20 \log_{10}(70/10) \\
&= 20 \log_{10} 7 \\
&= 20 \times 0.845 \\
&= 16.9 \, \text{dB}
\end{aligned}$$

This can be rounded off to 17 dB, because our input data is given to only two significant digits.

PROBLEM 7-2

If the rms AC input current to a circuit is 250 microamperes (μA) and the output is 100 milliamperes (mA), what is the current gain? Assume the input and output impedances are identical.

SOLUTION 7-2

First, we must convert the input and output current to the same units. Let's use milliamperes. Then $I_{in} = 0.250$ and $I_{out} = 100$, and we can use the above formula for current gain:

$$\begin{aligned}
\text{Gain (dB)} &= 20 \log_{10}(I_{out}/I_{in}) \\
&= 20 \log_{10}(100/0.250) \\
&= 20 \log_{10} 400 \\
&= 20 \times 2.602 \\
&= 52.04 \, \text{dB}
\end{aligned}$$

We can round this off to 52.0 dB, because our input data is given to three significant figures.

PROBLEM 7-3

Suppose an *RF attenuator* reduces the power applied to it by a factor of exactly 15 : 1. What is the power gain in decibels? The power loss?

SOLUTION 7-3

We are given a ratio here, and not actual power values. However, we can assign an input power value $P_{in} = 15$ units (whatever the unit might be) and $P_{out} = 1$ unit. Then, using the above formula for power gain:

$$\text{Gain (dB)} = 10 \log_{10} (P_{out}/P_{in})$$

$$= 10 \log_{10} (1/15)$$

$$= 10 \log_{10} 0.06667$$

$$= 10 \times (-1.176)$$

$$= -11.76\, \text{dB}$$

Because we're told the attenuation factor is exact, we can leave this figure as it is. An engineer would probably round it off to $-11.8\, \text{dB}$ or maybe even to $-12\, \text{dB}$. That is the gain of the attenuator. We can also say that the circuit produces $11.8\, \text{dB}$ or $12\, \text{dB}$ of power loss.

Basic Amplifier Circuits

In general, amplifiers must use active components, such as transistors or ICs. A simple AC transformer can increase the deliverable current or voltage if the input and output impedances differ. But it cannot produce an output signal that has more power than the input signal, nor can it increase the current or voltage if the input and output impedances are the same.

GENERIC BIPOLAR AMPLIFIER

A generic NPN bipolar-transistor amplifier is shown in Fig. 7-1. The input signal passes through a capacitor to the base. Resistors provide bias. In this amplifier, the capacitors must have values large enough to allow the AC signal to pass with ease. But they should not be much larger than the minimum necessary for this purpose. The ideal capacitance values depend on the design frequency of the amplifier, and also on the impedances at the input and output. In general, as the frequency and/or circuit impedance increase, less capacitance is needed. Resistor values depend on the input and output impedances.

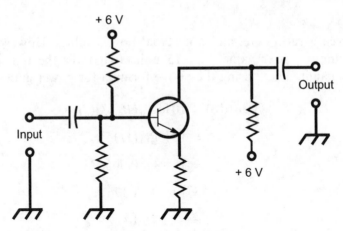

Fig. 7-1.   Generic NPN bipolar-transistor amplifier circuit.

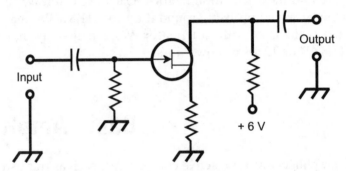

Fig. 7-2.   Generic N-channel FET amplifier circuit.

GENERIC FET AMPLIFIER

A generic N-channel JFET amplifier is shown in Fig. 7-2. Concerning the values of the capacitors, the same considerations apply here, as with the bipolar-transistor circuit. A JFET has high input impedance, and therefore the value of the input capacitor can be relatively small. If the device is a MOSFET, the input impedance is extremely high, and the input capacitance can be very small, sometimes less than 1 picofarad (pF). Resistor values depend on the input and output impedances.

GENERIC IC AMPLIFIER

A simple integrated-circuit (IC) amplifier is shown in Fig. 7-3. This circuit uses an operational amplifier (op amp). Op amps can perform various electronic functions as discussed in the previous chapter.

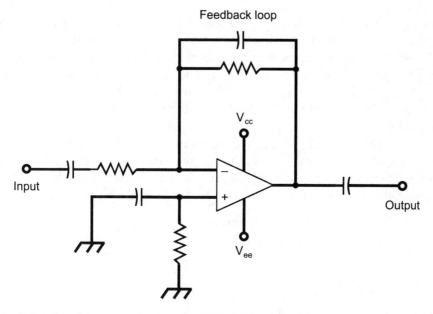

Fig. 7-3.   Generic op-amp circuit, using RC combinations to "tailor" the gain and the
frequency response.

The amplifier shown here is an *inverting amplifier*, meaning that the output
is 180° out of phase with the input. The RC combinations in the input and
the feedback loop are used to "tailor" the gain and the frequency response of
this circuit. In general, larger capacitances are used to favor low frequencies,
and small capacitances are used at higher frequencies. In the feedback loop,
low resistances produce low gain by increasing the *negative* (out-of-phase)
feedback, while large resistances produce high gain by reducing the negative
feedback.

Amplifier Classes

Amplifier circuits can be categorized as *class A*, *class AB*, *class B*, or *class C*.
Each class has unique characteristics and applications.

CLASS A

Weak-signal amplifiers, such as the kind used in a microphone preamplifier
or in the first stage of a sensitive radio receiver, are always *class-A amplifiers*.

This type of amplifier is linear. That means the shape of the output wave is a faithful, but magnified, reproduction of the shape of the input wave.

In order to obtain class-A operation with a bipolar transistor, the bias must be such that, with no signal input, the device operates near the middle of the straight-line portion of the collector current (I_C) versus base current (I_B) curve. This is shown for an NPN transistor in Fig. 7-4. For PNP, reverse the polarity signs.

With a JFET or MOSFET, the bias must be such that, with no signal input, the device is near the middle of the straight-line part of the drain current (I_D) versus gate voltage (E_G) curve. This is shown in Fig. 7-5 for an N-channel device. For P-channel, reverse the polarity signs.

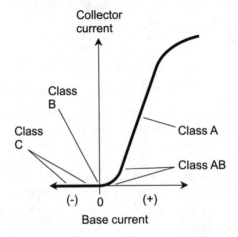

Fig. 7-4.   Classes of amplification for a bipolar transistor.

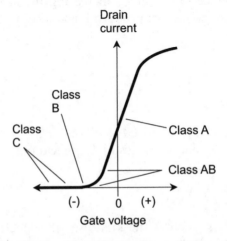

Fig. 7-5.   Classes of amplification for an N-channel JFET.

CLASS AB

When a bipolar transistor is biased slightly above the point where the no-signal base current becomes zero (cutoff), or when an FET is biased slightly above the point where the no-signal gate current becomes zero (pinchoff), the input signal drives the device into the nonlinear part of the operating curve. This is *class-AB operation*. Figure 7-4 shows this situation for an NPN bipolar transistor, and Fig. 7-5 shows it for an N-channel JFET.

In a class-AB amplifier, the input signal might cause the device to go into cutoff or pinchoff for a small part of the cycle. Whether or not this happens depends on the actual bias point, and also on the strength of the input signal. If the bipolar transistor or FET is never driven into cutoff/pinchoff during any part of the signal cycle, the circuit is a *class-AB$_1$ amplifier*. If the device goes into cutoff/pinchoff for any part of the cycle, the circuit is a *class-AB$_2$ amplifier*.

In any class-AB amplifier, the output waveform differs in shape from the input waveform. But if the signal is modulated, such as in a voice radio transmitter, the data impressed on the signal will emerge undistorted anyway. That is to say, the *modulation envelope* will be undistorted even though the wave shape itself is distorted. Class-AB operation is commonly used in RF power amplifier (PA) systems.

CLASS B

When a bipolar transistor is biased exactly at cutoff, or when an FET is biased exactly at pinchoff, an amplifier is said to be working in class B. These operating points are labeled on the curves in Figs. 7-4 and 7-5. In a *class-B amplifier*, there is no collector or drain current when there is no signal. This saves energy compared with class-A and class-AB circuits. When there is an input signal, current flows in the device during exactly half of the cycle.

Sometimes two bipolar transistors or FETs are used in a class-AB or class-B circuit, one for the positive half of the cycle and the other for the negative half. In this way, distortion is eliminated. This is a *push-pull amplifier* (Fig. 7-6) and is commonly used in audio applications.

The class-B scheme can be used for RF power amplification. The output wave has a shape that is much different from that of the input wave, and this produces harmonics in addition to the signal at the fundamental frequency. This can be a problem, but it can be overcome by a resonant circuit in the output. If the signal is modulated, the modulation envelope is not distorted.

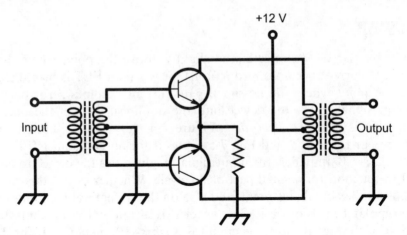

Fig. 7-6. A push-pull amplifier.

CLASS C

A bipolar transistor or FET can be biased past cutoff or pinchoff, and it can work as a PA if the drive is sufficient to overcome the bias and cause the device to conduct during part of the cycle. This is a *class-C amplifier*. Approximate operating points for class C are labeled in Figs. 7-4 and 7-5.

A class-C RF PA is nonlinear for signal envelopes in which the amplitude varies over a continuous range. An example is a standard *amplitude-modulation* (AM) signal. The class-C circuit will work properly only for a signal whose amplitude is constant, or else has only two states (called on/off, high/low, or mark/space). Continuous-wave (CW) *radiotelegraphy, radio-teletype* (RTTY), and *frequency modulation* (FM) are examples of such signals.

A class-C RF PA needs substantial driving power in order to overcome the cutoff or pinchoff bias that is applied to its base or gate. When properly operated, however, it can work with high efficiency and is still used in some broadcast transmitters.

Efficiency and Drive

In a PA, *efficiency* is the ratio of the useful power output to the total power input. High efficiency ensures minimal generation of heat, and also maximum component life.

DC POWER INPUT

The *DC power input* (P_{in}), in watts, to a bipolar-transistor amplifier circuit is the product of the collector current (I_C) in amperes and the collector voltage (E_C) in volts. For an FET, the DC power input is the product of the drain current (I_D) and the drain voltage (E_D). Mathematically:

$$P_{in} = E_C I_C$$

for bipolar-transistor power amplifiers, and

$$P_{in} = E_D I_D$$

for FET power amplifiers.

In some circuits, the DC power input is considerable even when there is no signal applied. In class A, for example, the average DC power input is constant whether there is an input signal or not. In class AB_1 or AB_2, there is a small DC power input with no signal input, and significant DC power input with the application of a signal. In class B and class C, there is zero DC power input when there is no input signal, and significant DC power input with the application of a signal.

SIGNAL POWER OUTPUT

When there is no signal input to an amplifier, there is no signal at the output, and therefore the *signal power output* (P_{out}) is zero. This is true no matter what the class of amplification. The greater the signal input, in general, the greater the power output of a power amplifier, up to a certain point.

The signal power output from an amplifier cannot be directly measured using DC instruments. Specialized AC wattmeters are necessary.

DEFINITION OF EFFICIENCY

The *efficiency* (eff) of a PA is the ratio of the signal power output to the DC power input:

$$\text{eff} = P_{out}/P_{in}$$

This ratio is always between 0 and 1. Efficiency is often expressed as a percentage between 0% and 100% (eff$_\%$), so the formula becomes

$$\text{eff}_\% = (100 P_{\text{out}} / P_{\text{in}})\%$$

EFFICIENCY VERSUS CLASS

The efficiency of a class-A amplifier is low, ranging from 25% to 40%, depending on the nature of the input signal and the type of bipolar or field-effect transistor used. If the input signal is weak, such as is the case at the antenna input of a radio receiver, the power-amplification efficiency of a class-A circuit is near 0%. But in that application, efficiency is not of primary importance. High gain and low *noise figure* (internally generated circuit noise) are vastly more important.

A class-AB$_1$ RF PA is 35% to 45% efficient. The efficiency of a class-AB$_2$ RF PA can approach 60%. Class-B amplifiers are 50% to 65% efficient, and class-C RF PA systems can be up to 80% efficient.

DRIVE AND OVERDRIVE

In theory, class-A and class-AB$_1$ power amplifiers draw no power from a signal source in order to produce useful power output. Class-AB$_2$ amplifiers need some driving power to produce output. Class-B amplifiers require more drive than class-AB$_2$, and class-C amplifiers need still more. Whatever class of amplification is used, it is important that the driving signal not be too strong. If *overdrive* takes place, problems occur.

In Fig. 7-7, the output signal waveshapes for power amplifiers are shown in various situations, assuming that the input is a perfect sine wave. In drawings A and B, the amplifier operates in class-A or in class-B push-pull. The waveform at A shows the signal output from a properly driven amplifier, and the waveform at B shows the signal output from an overdriven amplifier. Drawings C and D show the output signal waveforms from properly driven and overdriven class-B RF PAs, respectively. Drawings E and F show the output waveforms from properly driven and overdriven class-C RF PAs, respectively.

Note the *flat topping* that occurs with overdrive. This can cause excessive harmonic emission, modulation-envelope distortion, and reduced efficiency.

PROBLEM 7-4
Suppose an RF PA operates with a collector current of 800 mA and a collector voltage of 12 V. The RF power output is 4.8 W. What is the efficiency of this amplifier in percent?

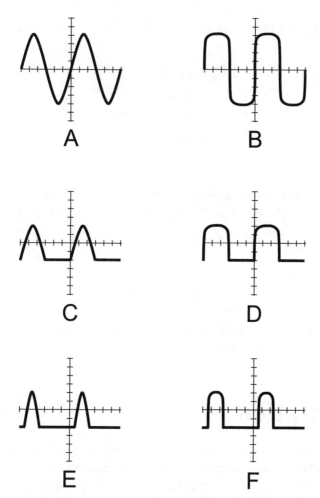

Fig. 7-7. Output waveforms of power amplifiers: properly driven class-A (shown at A); overdriven class-A (shown at B); properly driven class-B (shown at C); overdriven class-B (shown at D); properly driven class-C (shown at E); overdriven class-C (shown at F).

SOLUTION 7-4

First, determine the DC power input. The collector current, I_C, is 0.800 A and the collector voltage, E_C, is 12 V. Therefore:

$$P_{in} = E_C I_C$$

$$= 12 \times 0.800$$

$$= 9.6\,\text{W}$$

The RF power output P_{out}, is given as 4.8 W. Therefore, the efficiency in percent, $eff_\%$, is:

$$eff_\% = (100P_{out}/P_{in})\%$$
$$= (100 \times 4.8/9.6)\%$$
$$= (100 \times 0.5)\%$$
$$= 50\%$$

PROBLEM 7-5

Suppose an amplifier is supplied with a sine-wave input signal, with a waveform that looks like Fig. 7-7A. The output signal waveform looks like Fig. 7-7D. What can we say about the effect this circuit has on the input signal waveform?

SOLUTION 7-5

The negative parts of every cycle are cut off. This is the equivalent of half-wave rectification. In addition, the output waveform is distorted. This is not only because of the half-wave rectification, but there is a change in the shapes of the portions of the cycle that remain.

Audio Amplification

High-fidelity (hi-fi) audio amplifiers work from a few hertz up to approximately 100 kHz. Audio amplifiers for voice communications cover about 300 Hz to 3 kHz. In digital communications, audio amplifiers work over a very narrow range of frequencies.

FREQUENCY RESPONSE

Hi-fi amplifiers are equipped with resistor-capacitor (RC) *tone controls* that tailor the frequency response. The simplest tone control uses a single rotatable rotary or slide potentiometer. More sophisticated systems have separate controls, one for "bass" and the other for "treble." The most advanced systems have *graphic equalizers*, sets of several controls that affect the amplifier gain over various frequency spans.

Gain-versus-frequency curves for three hypothetical audio amplifiers are shown in Fig. 7-8. At A, a wideband, flat curve is illustrated. This is typical of hi-fi system amplifiers. At B, a voice communications response is

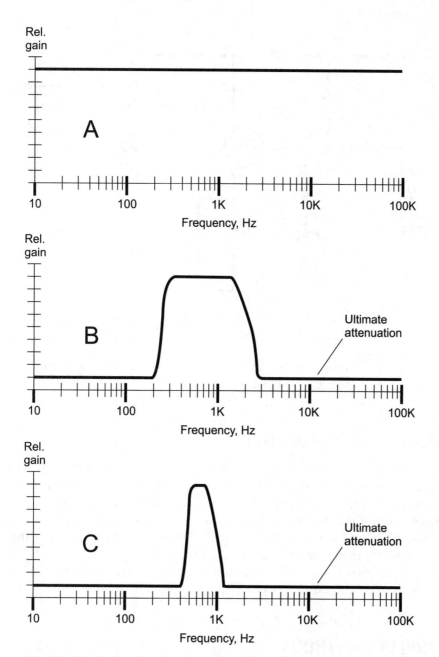

Fig. 7-8.   Audio amplifier frequency response curves: typical hi-fi (A), voice communications (B), and digital communications (C).

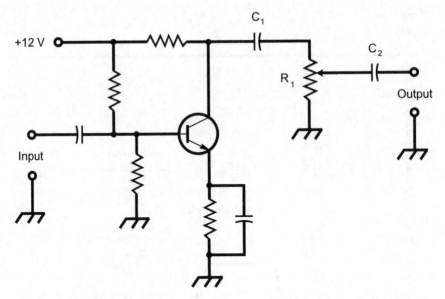

Fig. 7-9. A simple audio-amplifier volume control. The gain is adjusted with R_1.

shown. At C, a narrowband response curve, typical of audio amplifiers in digital wireless receivers, is illustrated.

VOLUME (GAIN) CONTROL

Audio amplifier systems usually consist of several stages. In an early stage, a *volume control* is used. An example is shown in Fig. 7-9. The volume control is resistor R_1, a potentiometer. The AC output signal passes through C_1 and appears across R_1. The wiper (indicated by the arrow) of the potentiometer "picks off" more or less of the AC output signal. Capacitor C_2 isolates the potentiometer from the DC bias of the following stage.

Volume control is usually done in a stage where the audio power level is low. This allows the use of a potentiometer rated for about 1 W.

COUPLING METHODS

The term *coupling* refers to the method used to get a signal from one stage in a complex circuit to the next, such as from a low-power amplifier to a higher-power amplifier. An example of *transformer coupling* is shown in Fig. 7-10. Capacitors C_1 and C_2 keep one end of the transformer primary

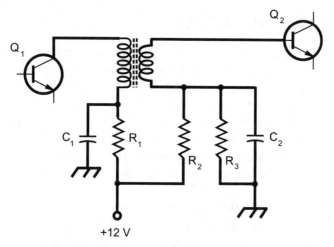

Fig. 7-10.   An example of transformer coupling.

and secondary at signal ground. Resistor R_1 limits the current through the first transistor, Q_1. Resistors R_2 and R_3 provide the proper base bias for transistor Q_2. This scheme provides optimum signal transfer between amplifier stages with minimum signal loss, because of the impedance-matching ability of the transformer.

The output impedance of Q_1 is perfectly matched to the input impedance of Q_2 by using a transformer with the correct turns ratio.

In some amplifier systems, capacitors are added across the primary and/or secondary of the transformer. This results in *resonance* at a frequency determined by the capacitance and the transformer winding inductance. If the set of amplifiers is intended for a single frequency, *tuned-circuit coupling* enhances the efficiency. When tuned-circuit coupling is used, care must be taken to ensure that the amplifier chain does not oscillate at the resonant frequency.

RF Amplification

The *RF spectrum* begins at a few kilohertz (there is disagreement in the literature about the exact lower limit), and extends to over 300 GHz. At the low-frequency end of this range, RF amplifiers resemble AF amplifiers. As the frequency increases, amplifier designs and characteristics change.

WEAK-SIGNAL AMPLIFIERS

The *front end*, or first amplifying stage, of a radio receiver requires the most sensitive possible amplifier. Sensitivity is determined by two factors: *gain* and *noise figure*. Gain is measured in decibels (dB). Noise figure is a measure of how well a circuit or system can amplify desired signals without introducing unwanted noise. All bipolar transistors or FETs create some noise – energy whose frequencies span a broad range – because of the motions of the charge carriers within the materials. In general, FETs produce less noise than bipolar transistors. In the output of a receiver, noise sounds like a hiss or roar.

Weak-signal amplifiers almost always use resonant circuits. This optimizes the amplification at the desired frequency, while helping to cut out noise from unwanted frequencies. A tuned weak-signal RF amplifier is shown in Fig. 7-11. This circuit uses a device called a *gallium arsenide FET* (abbreviated GaAsFET and pronounced "GAS-fet") and is designed for approximately 10 MHz. At higher frequencies, the inductances and capacitances are smaller; at lower frequencies, the values are larger.

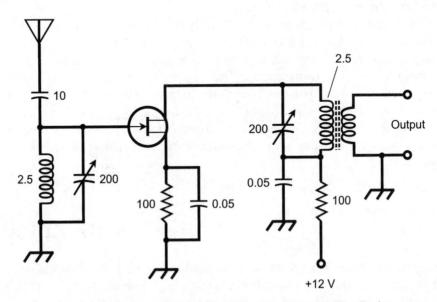

Fig. 7-11.   A tuned weak-signal RF amplifier for use at about 10 MHz. Resistances are in ohms. Capacitances are in microfarads (μF) if less than 1, and in picofarads (pF) if greater than 1. Inductances are in microhenrys (μH).

BROADBAND RF POWER AMPLIFIERS

The main advantage of a *broadband RF PA* is ease of operation, because the circuit does not require tuning. The operator need not worry about critical adjustments. However, broadband RF PAs are slightly less efficient than their tuned counterparts.

A broadband PA amplifies any signal whose frequency is within its design range, whether or not that signal is intended for transmission. If some circuit in a transmitter has too much feedback, causing *oscillation* at a frequency other than the intended signal frequency, and if this undesired signal falls within the design frequency range of a broadband PA, the unwanted signal is amplified along with the desired one. The result is a phenomenon called *spurious emissions*.

A typical broadband PA circuit is diagrammed schematically in Fig. 7-12. The amplifying device is an NPN power transistor. The transformers are a critical part of this circuit. They must be designed to function efficiently over a wide range of frequencies. This circuit will work well from approximately 1 MHz through 30 MHz.

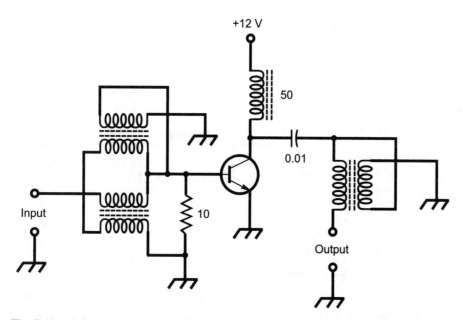

Fig. 7-12.   A broadband RF power amplifier, capable of producing a few watts output. Resistances are in ohms. Capacitances are in microfarads (μF) if less than 1, and in picofarads (pF) if greater than 1. Inductances are in microhenrys (μH).

TUNED RF POWER AMPLIFIERS

A *tuned RF PA* offers improved efficiency compared with broadband designs. Also, the tuning helps to reduce the probability that spurious signals will be amplified and transmitted over the air. A tuned RF PA can deliver output into circuits or devices having a wide range of impedances. In addition to a *tuning control*, or resonant circuit that adjusts the output of the amplifier to the operating frequency, there is a *loading control* that optimizes the signal transfer between the amplifier and the load.

The main drawbacks of a tuned PA are the facts that adjustment can be time-consuming, and improper adjustment can result in damage to the amplifying device (bipolar transistor or FET). If the tuning and/or loading controls are improperly set, the efficiency will be near zero while the DC collector or drain power input is high.

A tuned RF PA, providing useful power output at approximately 10 MHz, is shown in Fig. 7-13. The tuning and loading controls must be adjusted for maximum RF power output as indicated on a wattmeter in the feed line going to the load.

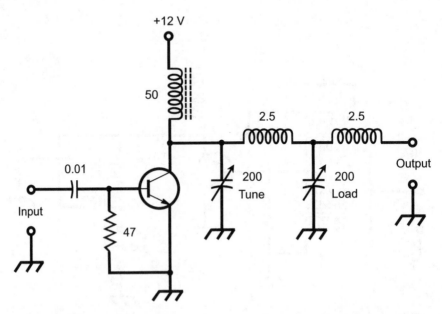

Fig. 7-13.   A tuned RF power amplifier, capable of producing a few watts output. Resistances are in ohms. Capacitances are in microfarads (μF) if less than 1, and in picofarads (pF) if greater than 1. Inductances are in microhenrys (μH).

PROBLEM 7-6
What do the dashed lines represent in the transformer and inductor symbols in Fig. 7-12?

SOLUTION 7-6
Check the schematic symbols appendix when you're not sure what a symbol means, or when you need to be sure you use the correct symbol when drawing a diagram. From the appendix, it is apparent that dashed lines represent powdered-iron cores.

PROBLEM 7-7
What is the purpose of a powdered-iron core in a transformer or inductor? Why are these core types specified in the diagram of Fig. 7-12, rather than air-core types?

SOLUTION 7-7
When a transformer or inductor has a powdered-iron core, the inductance of each coil winding is increased, compared with its inductance if the core is air. This makes it possible to use transformers and coils having fewer windings than are necessary if the cores are air. In addition, powdered iron concentrates most of the magnetic flux in the core material, so less of it appears outside the coil where it can cause interaction with other components. Air-core transformers and coils can be (and sometimes are) used in RF applications, where the required inductances are fairly small. But air-core coils don't work well in AF applications, because it's impossible to get enough inductance.

PROBLEM 7-8
How would the diagram of Fig. 7-12 differ if solid or laminated iron cores were used in the transformers and inductors? Why are powdered-iron cores used in the circuit shown by Fig. 7-12, rather than solid or laminated iron cores?

SOLUTION 7-8
Solid or laminated iron cores are indicated by pairs of solid, rather than dashed, parallel lines alongside or between the windings in an inductor. Transformers and coils having cores of this type are used in power supplies at the 60-Hz utility line frequency. Solid or laminated iron cores do not work well at AF or RF, because circulating currents arise within them. If solid or laminated cores are used in transformers and inductors at AF or RF, the circulating currents in the iron heat up the cores, and this takes away from the ability of the component to behave properly as a transformer or inductor.

204

PART 2   Wired Electronics

Quiz

Refer to the text in this chapter if necessary. Answers are in the back of the book.

1. The output signal waveform of a class C amplifier
    (a) is different from the input signal waveform.
    (b) is identical with the input signal waveform.
    (c) is always in phase with the input signal waveform.
    (d) is biased at cutoff or pinchoff.

2. A push-pull amplifier makes use of
    (a) nonlinear devices.
    (b) two bipolar transistors or FETs.
    (c) two rectifier diodes.
    (d) coils with laminated iron cores.

3. An important factor to consider in the design of a weak-signal amplifier is
    (a) the efficiency.
    (b) the linearity.
    (c) the noise figure.
    (d) the power output.

4. Suppose the input and output impedances of an amplifier are the same, and the signal voltage gain is 20 dB. This corresponds to an output-to-input voltage ratio of
    (a) 20 to 1.
    (b) 14.14 to 1.
    (c) 10 to 1.
    (d) 7.071 to 1.

5. If the output of an amplifier is 180° out of phase with the input, the amplifier
    (a) is inverting.
    (b) causes no phase shift.
    (c) is nonlinear.
    (d) has negative gain.

6. An advantage of class-C operation over class-A operation in an RF power amplifier, when used for digital signals, is the fact that
    (a) the gain is usually lower.
    (b) the noise figure is usually higher.

(c) the efficiency is usually higher.
(d) Stop right there! Class C amplifiers never make good RF power amplifiers.

7. Suppose a wireless transmitter has a broadband RF PA feeding the antenna, and unwanted spurious signals are radiated. These spurious signals can be reduced by
(a) switching to a tuned RF PA.
(b) switching to a class C RF PA.
(c) switching from a bipolar transistor PA to one that uses an FET.
(d) increasing the noise figure of the PA.

8. When the output waveform from an amplifier is a perfect, although magnified, replica of the input waveform, the amplifier
(a) has a high noise ratio.
(b) is operating in class C.
(c) has good linearity.
(d) has high distortion.

9. Overdriving a PA can cause
(a) flat-topping of the output waveform.
(b) an improved noise figure.
(c) excessive gain.
(d) improved efficiency.

10. Suppose a circuit cuts the signal power in half. This is a power gain of
(a) 6 dB.
(b) −6 dB.
(c) 3 dB.
(d) −3 dB.

CHAPTER 8

Signal Oscillators

An *oscillator* is essentially an amplifier with positive feedback that causes it to generate a signal of its own. Some oscillators work at AF, and others are intended to produce RF signals. Most oscillators generate sine waves. Some emit square waves, sawtooth waves, or other waveshapes.

RF Oscillators

In order for an amplifying circuit to oscillate, the gain must be high, the feedback must be positive, and the coupling from output to input must be good. Common-emitter and common-source circuits work well for oscillator design. Common-base and common-gate circuits can be made to oscillate, but they are more often used as power amplifiers.

FEEDBACK

The frequency of an RF oscillator can be controlled by means of a tuned, or resonant, circuit in the *feedback loop* (Fig. 8-1). This resonant circuit is usually an inductance-capacitance (LC) or resistance-capacitance (RC) combination. The LC scheme is common at RF; the RC method is often used in audio oscillators.

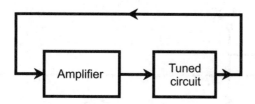

Fig. 8-1. The basic concept of an oscillator.

The tuned circuit in an oscillator has low loss at a single frequency, but high loss at other frequencies. The result is that the oscillation takes place at a predictable and stable frequency, determined by the inductance and capacitance, or by the resistance and capacitance.

ARMSTRONG CIRCUIT

A common-emitter or common-source amplifier can be made to oscillate by feeding the output back to the input through a transformer. The signal at the collector or drain is out of phase with the signal at the base or gate, so it's necessary to invert the fed-back signal phase to produce positive feedback. This can be done by connecting the transformer secondary to the base or gate "backwards." The schematic diagram of Fig. 8-2A shows a common-emitter NPN bipolar-transistor amplifier whose collector circuit is coupled to the emitter circuit through a transformer. At B, the equivalent N-channel JFET circuit is shown. These circuits are called *Armstrong oscillators*.

The oscillation frequency is determined by a capacitor across either the primary or the secondary winding of the transformer. The inductance of the winding, along with the capacitance connected in parallel with it, forms a resonant circuit. The fundamental frequency of oscillation is determined according to the formula

$$f = 1/[2\pi(LC)^{1/2}]$$

where f is the frequency in megahertz, L is the inductance of the transformer winding in microhenrys, and C is the value of the parallel capacitor in microfarads. Alternatively, for low-frequency applications, f can be expressed in hertz, L in henrys, and C in farads.

HARTLEY CIRCUIT

A method of obtaining controlled feedback at RF is shown in Fig. 8-3. At A, an NPN bipolar transistor is used; at B, an N-channel JFET is employed.

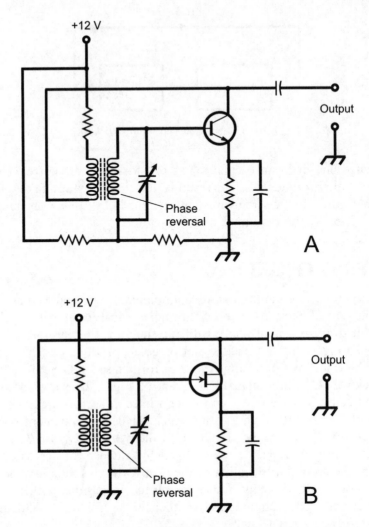

Fig. 8-2.   Armstrong oscillators. At A, an NPN bipolar transistor is used. At B, an N-channel JFET is used.

These circuits use a single coil with a tap on the windings to provide the feedback. A variable capacitor in parallel with the coil determines the oscillating frequency, and allows for frequency adjustment. These are examples of *Hartley oscillators*.

The Hartley circuit employs about 25% of its amplified power to produce feedback. The rest of the power is available as signal output. Oscillators rarely produce more than a fraction of a watt of output power. If more power is needed, the signal amplitude must be boosted by amplifier circuits following the oscillator.

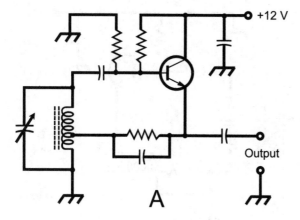

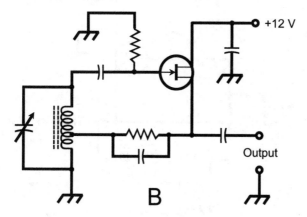

Fig. 8-3.   Hartley oscillators. At A, an NPN bipolar transistor is used. At B, an N-channel JFET is used.

In a Hartley oscillator, it is important to use only the minimum amount of feedback necessary to obtain oscillation. The amount of feedback is controlled by the position of the coil tap, and occurs between the emitter or source circuit and the base or gate circuit.

COLPITTS CIRCUIT

Another way to provide RF feedback is to tap the capacitance instead of the inductance in a tuned circuit. In Fig. 8-4, NPN bipolar (at A) and N-channel JFET (at B) *Colpitts oscillator* circuits are diagrammed.

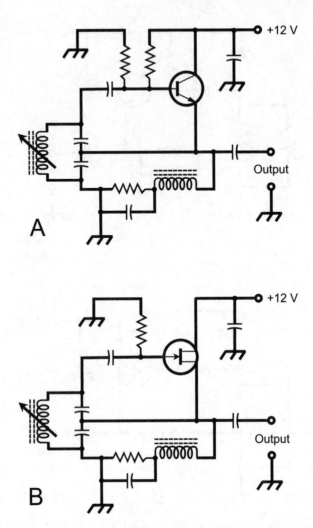

Fig. 8-4. Colpitts oscillators. At A, an NPN bipolar transistor is used. At B, an N-channel JFET is used.

The amount of feedback is controlled by the ratio of capacitances. The oscillation frequency depends on the net series capacitance C, and also on the inductance L of the variable coil. The general formula for resonant frequency applies (see above). If the values of the two series capacitors, connected across the variable inductor, are C_1 and C_2, then the net capacitance C is

$$C = C_1 C_2 / (C_1 + C_2)$$

All capacitances must be in the same units for this formula to work.

The inductance in the Colpitts circuit can be varied to obtain frequency adjustment. This is usually done by sliding a powdered-iron *slug* in and out of a coil wound on a hollow cylindrical form. A Colpitts circuit offers exceptional stability and reliability when properly designed. As with the Hartley circuit, the feedback should be kept to the minimum necessary to sustain oscillation.

CLAPP CIRCUIT

A variation of the Colpitts oscillator employs a series-resonant tuned circuit. A schematic diagram of an NPN bipolar-transistor *Clapp oscillator* is shown in Fig. 8-5A; the equivalent N-channel JFET circuit is shown at B.

The Clapp oscillator offers excellent stability at RF. Its frequency does not fluctuate appreciably when high-quality components are used. The Clapp is a reliable circuit, and it is easy to make it oscillate. This circuit allows the use of a variable capacitor for frequency control.

In the Hartley, Colpitts, and Clapp configurations, the output is usually taken from the emitter or source to optimize stability. To prevent the output signal from being short-circuited to ground, an RF choke is connected in series with the emitter or source in the Colpitts and Clapp circuits. Typical inductance values for RF chokes are $100\,\mu H$ at high frequencies such as $15\,MHz$, and $10\,mH$ at low frequencies such as $150\,kHz$.

DIODE OSCILLATOR

At ultra-high and microwave frequencies, certain diodes can be used as oscillators. Examples include Gunn, IMPATT, and tunnel diodes. We learned about these types of diodes in Chapter 5.

Oscillator Stability

With respect to oscillators, the term *stability* has two meanings: (1) constancy of frequency, and (2) reliability of performance.

FREQUENCY STABILITY

Hartley, Colpitts, and Clapp oscillators allow for frequency adjustment using variable capacitors or variable inductors. Electronic component values

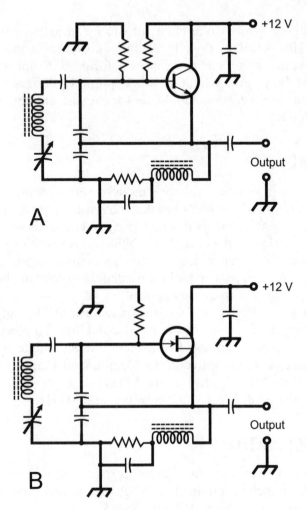

Fig. 8-5.   Clapp oscillators. At A, an NPN bipolar transistor is used. At B, an N-channel JFET is used.

are commonly affected by changes in the temperature. In any oscillator, it is crucial that the components maintain constant values, as much as possible, under all anticipated conditions.

Some types of capacitors maintain their values better than others as the temperature rises or falls. *Polystyrene capacitors* are excellent in this respect. *Silver-mica capacitors* work well when polystyrene units cannot be found. *Air-variable capacitors* are excellent in this respect (assuming they are left at the same capacitance settings, of course!).

Inductors are most temperature-stable when they have air cores. Air-core coils should be wound, when possible, from stiff wire with strips of plastic to keep the windings in place. Some air-core coils are wound on hollow cylindrical forms made of ceramic or phenolic material. Ferromagnetic core materials such as powdered iron work all right for many applications, but the *core permeability* is affected by temperature. This changes the inductance, in turn affecting the oscillator frequency.

Problems with oscillator frequency stability have been largely overcome in recent years by the evolution of *frequency synthesizers*. These are described later in this chapter.

RELIABILITY

An oscillator should begin functioning as soon as the power supply is switched on. It should keep oscillating under all normal conditions, not quitting if the load changes or the temperature suddenly rises or falls. The failure of an oscillator can cause an entire communications station to go down.

The oscillator circuits described in this chapter are those that engineers have found, through decades of trial and error, to work well. Even so, when an oscillator is built and put to use, *debugging* is often necessary. If two oscillators are built according to the same schematic diagram, with the same component types and values, and in the same physical arrangement, one circuit might be stable while the other is not. The usual reason is a difference in the quality or tolerance of one or more components. (Some frustrated engineers have been known to blame *gremlins*, which are tiny, imaginary monsters. According to the lore, the sole reason for these creatures' existence is to do mischief to electronic, computer, and mechanical systems of all kinds, especially when no one is looking.)

Most oscillators are designed to operate with high load impedances. If the load impedance is low, the load will "try" to draw power from the oscillator. Under these conditions, even a well-designed oscillator can become unstable. Signal power should be delivered by an amplifier circuit following an oscillator, not by the oscillator itself.

PROBLEM 8-1
In the Clapp oscillator circuit of Fig. 8-5A, how is the oscillation frequency related to the capacitance of the variable capacitor?

SOLUTION 8-1
As the capacitance of the variable capacitor increases, the capacitance of the set of capacitors (there are four of them in series) across the coil also increases. Remember the formula for the resonant frequency of a tuned

circuit. The frequency varies inversely in proportion to the square root of the capacitance. Thus, as the capacitance of the variable capacitor increases, the oscillation frequency goes down. As the capacitance of the variable capacitor decreases, the oscillation frequency goes up.

PROBLEM 8-2
In the Colpitts oscillator circuit of Fig. 8-4B, what is the purpose of the fixed capacitor between the tuned circuit and the gate of the FET?

SOLUTION 8-2
This capacitor allows the DC bias at the gate to be controlled by a resistor between the gate and ground. If the fixed capacitor were not there, the gate would be shorted to ground for DC through the coil in the tuned circuit.

Crystal-Controlled Oscillators

Quartz crystals, also called *piezoelectric crystals*, can be used in place of tuned LC circuits in RF oscillators, if it is not necessary to change the frequency often. *Crystal-controlled oscillators* offer frequency stability superior to that of LC-tuned oscillators. There are several ways that crystals can be connected in bipolar-transistor or FET circuits to obtain oscillation.

PIERCE CIRCUIT

The *Pierce oscillator* is a common crystal-controlled oscillator (Fig. 8-6). At A, an NPN bipolar transistor is used. At B, an N-channel JFET is employed.

The frequency of a Pierce oscillator can be varied by about plus-or-minus one-tenth of one percent ($\pm 0.1\%$) by means of an inductor in series with the crystal, or by means of a capacitor in parallel with the crystal. The frequency is determined mainly by the thickness of the crystal, and by the angle at which it is cut from the original quartz rock.

Crystals change in frequency as the temperature changes. They are more stable than LC circuits, most of the time. Some crystal oscillators are housed in temperature-controlled chambers called *crystal ovens*. They maintain their frequency so precisely that they are used as standards against which other oscillators are calibrated. The accuracy can be within a few hertz at working frequencies of several megahertz.

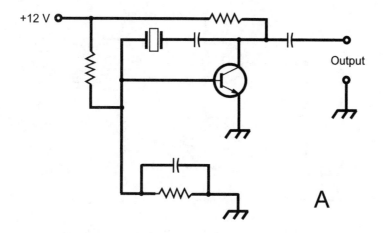

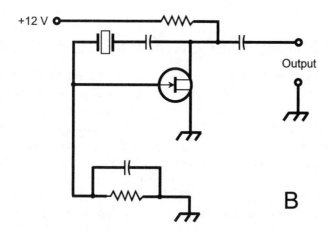

Fig. 8-6.   Pierce oscillators. At A, an NPN bipolar transistor is used. At B, an N-channel JFET is used.

REINARTZ CIRCUIT

A *Reinartz crystal oscillator* is characterized by high efficiency and minimal output at harmonic frequencies. The Reinartz configuration can be used with FETs or bipolar transistors.

Figure 8-7 is a schematic diagram of a Reinartz oscillator that employs an N-channel JFET. A resonant LC circuit is inserted in the source line. This resonant circuit is tuned to approximately half the crystal frequency. This allows the circuit to oscillate at a low level of crystal current. The LC

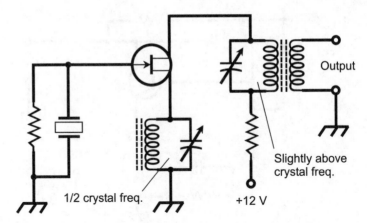

Fig. 8-7. A Reinartz crystal oscillator using an N-channel JFET.

circuit in the drain line is tuned to a frequency slightly above that of the crystal.

VARIABLE-FREQUENCY CRYSTAL OSCILLATOR (VXO)

The frequency of a *variable-frequency crystal oscillator* (VXO) can be varied slightly with the addition of a reactance in series or parallel with the crystal. This is called *frequency trimming*. VXOs are sometimes used in transmitters or transceivers to obtain operation over a small range of frequencies.

Examples of bipolar-transistor VXO circuits are shown in Fig. 8-8. At A, a variable inductor is connected in series with the crystal. At B, a variable capacitor is connected in parallel with the crystal.

The main advantage of the VXO over an oscillator whose frequency is determined by inductors and capacitors is excellent frequency stability. The main disadvantage of the VXO is limited frequency coverage. In most practical cases, the frequency cannot be varied by more than ±0.1% of the operating frequency without loss of stability. At 10 MHz, for example, that is a range of only ±10 kHz.

PROBLEM 8-3

In the crystal-controlled oscillator circuit shown in Fig. 8-8B, what would happen if the capacitor between the base and emitter were to short out?

SOLUTION 8-3

This would place the base and emitter at the same voltage, and would also short out any input signal to the transistor. The transistor acts as an amplifying device. If it is deprived of an input signal, it is also prevented

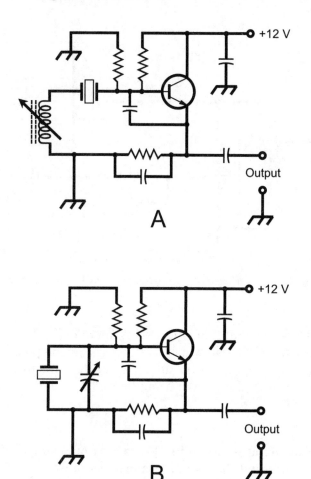

Figure 8-8.   Bipolar-transistor VXO circuits. At A, inductive frequency trimming; at B, capacitive frequency trimming.

from producing any output. Thus, shorting of this capacitor would cause failure of the oscillator.

VOLTAGE-CONTROLLED OSCILLATOR (VCO)

The frequency of a VFO can be adjusted by means of a *varactor diode* in the tuned LC circuit. Hartley and Clapp oscillator circuits are easily adapted to varactor frequency control. The varactor is isolated for DC by *blocking capacitors*. The frequency is adjusted by applying a variable DC voltage to the varactor.

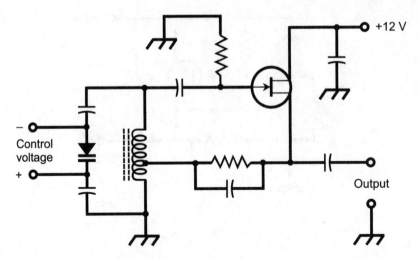

Fig. 8-9.   A Hartley voltage-controlled oscillator (VCO) using an N-channel JFET.

The schematic diagram of Fig. 8-9 shows a JFET Hartley *voltage-controlled oscillator* (VCO) in which a varactor is used in place of the variable capacitor. Note the capacitors on either side of the varactor. These keep the DC varactor control voltage from being shorted to ground, either directly or through the coil.

PROBLEM 8-4
In the oscillator circuit shown in Fig. 8-9, what would happen if the capacitor between the source and the output terminal were to open?

SOLUTION 8-4
This would not stop the circuit from oscillating, but it would prevent the signal from reaching the output terminal. The net effect on circuits following the oscillator would be the same as if the oscillator had failed.

PLL FREQUENCY SYNTHESIZER

An oscillator that combines the flexibility of a VFO with the stability of a crystal oscillator is known as a *phase-locked-loop (PLL) frequency synthesizer*. This scheme is extensively used in radio transmitters and receivers.

In a frequency synthesizer, the output of a VCO is passed through a *programmable divider*, which is a digital circuit that can divide the VCO frequency by almost any whole-number factor. The output frequency of the programmable divider is locked, by means of a *phase comparator*, to the

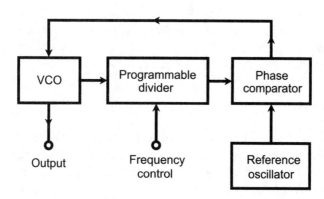

Fig. 8-10. Block diagram of a phase-locked-loop (PLL) frequency synthesizer.

signal from a crystal-controlled *reference oscillator*. As long as the output from the programmable divider is exactly on the reference-oscillator frequency, the two signals are in phase, and the output of the comparator is zero volts DC. If the VCO frequency "tries" to drift, the output frequency of the programmable divider also "tries" to drift (at a different rate). Even a tiny frequency change causes the phase comparator to produce a *DC error voltage*. This voltage is either positive or negative, depending on whether the VCO has drifted higher or lower in frequency. The DC error voltage is applied to a varactor in the VCO, causing the VCO frequency to change in a direction opposite to that of the drift. This forms a *DC feedback loop* that maintains the VCO frequency at a precise multiple of the reference-oscillator frequency, that multiple having been chosen by the programmable divider. Think of the process as similar to what takes place in your mind and body when you steer a car along a straight highway.

The key to the stability of the PLL frequency synthesizer lies in the fact that the reference oscillator is crystal-controlled. Figure 8-10 is a block diagram of a PLL frequency synthesizer. The frequency stability can be enhanced by using, as the reference oscillator, an amplified signal from the National Institute of Standards and Technology (NIST). These signals are transmitted on the shortwave bands by radio stations WWV or WWVH at 5, 10, or 15 MHz.

AF Oscillators

Audio-frequency (AF) oscillators are used in doorbells, ambulance sirens, electronic games, personal computers, and toys that play musical tunes.

All AF oscillators consist of AF amplifiers with positive feedback. Most AF oscillators use RC circuits to determine the frequency. Once in a while, you'll see an AF oscillator that uses an LC circuit.

WAVEFORMS

At RF, oscillators are usually designed to produce a nearly perfect sine-wave output, representing energy at one well-defined frequency. But audio oscillators do not necessarily concentrate all their energy at a single frequency. There are applications, particularly in electronic music, where a sine-wave output signal is not desired.

Various musical instruments sound different, even when they play notes at the same pitch. This is because each instrument has its own unique waveform. An instrument's sound qualities can be reproduced using an AF oscillator whose waveform output matches that of the instrument. A computer can have a *musical-instrument digital interface* (*MIDI*) *player* that employs audio oscillators capable of duplicating the sounds of a large band or orchestra. This allows people to produce sophisticated electronic music using modest personal computers.

TWIN-T OSCILLATOR

A form of AF oscillator that is popular for general-purpose use is the *twin-T oscillator* (Fig. 8-11). The transistor on the right produces amplification, and also causes a 180° phase shift for the signal. The RC circuit (the combination of two resistors R and two capacitors C) produces a 180° phase shift. The transistor on the left acts as an emitter follower, stabilizing the operation of the oscillator. The signal "chases its own tail" through both transistors, and also through the RC circuit, producing oscillation.

The frequency is determined by the values of the resistors R and capacitors C. The output is "picked off" of the RC circuit. The circuit in this example uses NPN bipolar transistors. A similar circuit can be built using N-channel JFETs.

PROBLEM 8-5
In the circuit of Fig. 8-11, what would happen if the capacitor C on the left-hand side were to short out?

SOLUTION 8-5
This would upset the resonant circuitry and deprive the left-hand transistor of its input, interrupting the feedback loop and causing failure of the oscillator.

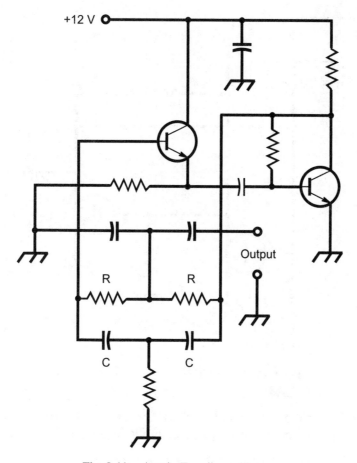

Fig. 8-11.   A twin-T audio oscillator.

PROBLEM 8-6

In the circuit of Fig. 8-11, what would happen if the connection between the right-hand transistor and ground were to open up?

SOLUTION 8-6

This would prevent the right-hand transistor from amplifying. This would cause failure of the oscillator because there would no longer be any amplification in the feedback loop.

MULTIVIBRATOR

A *multivibrator* uses two amplifier circuits, interconnected so the signal loops between them. N-channel JFETs can be connected as shown in Fig. 8-12.

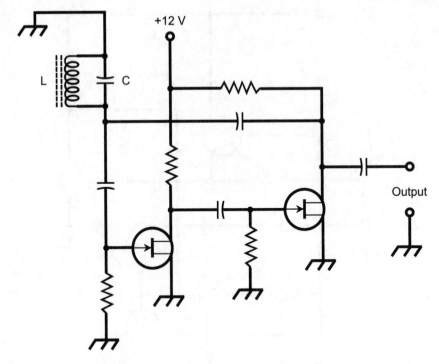

Fig. 8-12.   A multivibrator type audio oscillator.

Each JFET amplifies the signal in class-A, and reverses the phase by 180°. The frequency in this particular circuit is set by means of an LC resonant circuit.

The inductor in this example has a powdered-iron core shaped like a donut, called a *toroidal core*. This core shape allows for a much higher inductance, for a given number of coil turns, than a cylindrical core (also known as a *solenoidal core*). The value of the inductor is on the order of 0.5 H. The capacitance is chosen to obtain an audio tone at the frequency desired.

Quiz

Refer to the text in this chapter if necessary. Answers are in the back of the book.

1.   When an oscillator is said to be unstable, this could mean that
   (a) the frequency is absolutely constant.
   (b) the output is a sine wave.

    (c) the circuit does not oscillate reliably.

    (d) the circuit has positive feedback.

2.   A PLL frequency synthesizer

    (a) always produces a square wave.

    (b) has a stable signal frequency.

    (c) requires a Hartley circuit in order to work.

    (d) requires a Pierce circuit in order to work.

3.   In an Armstrong oscillator

    (a) the signal is fed from the output to the input through a transformer.

    (b) the signal is fed from the input to the output through a capacitor.

    (c) the signal output is taken from the base of a JFET.

    (d) a quartz crystal is used to determine the frequency.

4.   An oscillator that uses a tapped coil to obtain the feedback is called

    (a) a Hartley circuit.

    (b) a Pierce circuit.

    (c) a multivibrator.

    (d) a negative-feedback circuit.

5.   Suppose an oscillator uses an LC circuit in the feedback loop. If the inductance in this circuit is increased by a factor of 4 while the capacitance remains constant, what happens to the frequency of oscillation?

    (a) It goes down to 1/4 the previous frequency.

    (b) It goes down to 1/2 the previous frequency.

    (c) It doubles.

    (d) It goes up by a factor of 4.

6.   Examine Fig. 8-2B. Suppose that the tuned circuit marked "phase reversal" were to suddenly stop reversing the phase of the signals circulating through it. What would happen?

    (a) The oscillator output would increase, because the positive feedback would increase.

    (b) The circuit would stop oscillating, because the feedback would become negative instead of positive.

    (c) The frequency of oscillation would change because the effective inductance would change.

    (d) Nothing; the circuit would keep operating normally.

7. Which type of the following oscillator types uses a quartz crystal to determine the frequency?
    (a) The Hartley circuit.
    (b) The Clapp circuit.
    (c) The twin-T circuit.
    (d) The Pierce circuit.

8. Examine Fig. 8-3A. Suppose the variable capacitor is set to the center of its range, and then is readjusted so its capacitance reaches its maximum value. What happens to the output of the oscillator?
    (a) The frequency decreases.
    (b) The frequency increases.
    (c) The signal becomes a square wave.
    (d) The signal becomes a sawtooth wave.

9. In order to function properly, an oscillator must
    (a) use either JFETs or bipolar transistors.
    (b) must have a capacitor in the input circuit.
    (c) have a sufficient amount of feedback, but not too much.
    (d) produce a sine-wave output.

10. Examine Fig. 8-9. What does the "control voltage" control?
    (a) The signal frequency.
    (b) The signal strength.
    (c) The signal waveform.
    (d) The inductance in the LC circuit.

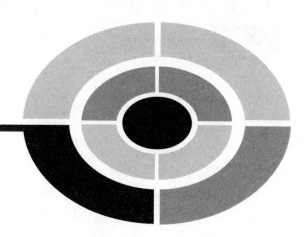

Test: Part Two

Do not refer to the text when taking this test. You may draw diagrams or use a calculator if necessary. A good score is at least 30 correct. Answers are in the back of the book. It's best to have a friend check your score the first time, so you won't memorize the answers if you want to take the test again.

1. A semiconductor diode can be used as any of the following except
   (a) a rectifier.
   (b) a signal envelope detector.
   (c) a high-voltage AC source.
   (d) a signal generator.
   (e) a high-speed switch.

2. What does Fig. Test2-1 most likely represent?
   (a) A characteristic curve for a diode.
   (b) A characteristic curve for a capacitor.
   (c) A characteristic curve for a resistor.
   (d) A characteristic curve for a JFET.
   (e) A characteristic curve for a bipolar transistor.

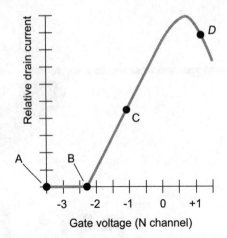

Fig. Test2-1. Illustration for Part Two Test Questions 2 through 5.

3. In Fig. Test2-1, which point, if any, is ideal for weak-signal amplification?
   (a) *A*
   (b) *B*
   (c) *C*
   (d) *D*
   (e) None of them.

4. Which, if any, of the points in Fig. Test2-1 is exactly at pinchoff?
   (a) *A*
   (b) *B*
   (c) *C*
   (d) *D*
   (e) None of them.

5. In Fig. Test2-1, which point, if any, represents improper bias for any common application?
   (a) *A*
   (b) *B*
   (c) *C*
   (d) *D*
   (e) All of the points represent proper bias.

6. When a bipolar transistor is biased exactly at cutoff (the point where the no-signal base current becomes zero), or when an FET is biased at pinchoff (the point where the no-signal gate current becomes zero), an amplifier is said to be working in
   (a) class A.

(b) class B.
(c) class C.
(d) class D.
(e) class E.

7. A common-emitter amplifier circuit always contains
    (a) a diode.
    (b) a JFET.
    (c) a bipolar transistor.
    (d) a transformer.
    (e) an RF choke.

8. A P-N junction will always conduct when
    (a) the forward voltage exceeds the forward breakover threshold.
    (b) the reverse voltage is zero.
    (c) the P-type electrode is connected to an antenna.
    (d) the N-type electrode is connected to ground.
    (e) any of the above situations exists.

9. For a circuit to oscillate, the gain must be high, the feedback must be
    positive, and
    (a) the input must be a sine wave.
    (b) the power supply voltage must be positive.
    (c) the device must use a bipolar JFET.
    (d) the inductive reactance must be low.
    (e) the coupling from output to input must be good.

10. An amplitude change of 1 dB is roughly equal to
    (a) a doubling of the signal power.
    (b) the largest change a listener or observer can detect if the change
        is expected.
    (c) the smallest change a listener or observer can detect if the change
        is expected.
    (d) the largest change a listener or observer can detect if the change
        is not expected.
    (e) A 10-fold increase in the signal voltage.

11. What function do the diodes perform in a power supply that pro-
    duces DC output when plugged into a utility wall outlet?
    (a) Oscillation.
    (b) Detection.
    (c) Switching.

(d) Current generation.

(e) Rectification.

12. Various musical instruments sound different, even when they play notes of the same frequency, because each instrument has its own unique

(a) wavelength.

(b) waveform.

(c) inductive reactance.

(d) capacitive reactance.

(e) resistance.

13. What is a disadvantage of a tuned RF power amplifier, compared with a broadband RF power amplifier?

(a) The tuned amplifier requires time-consuming adjustment when the frequency is changed significantly.

(b) The tuned amplifier cannot be adjusted to compensate for significant changes in frequency.

(c) The tuned amplifier allows more harmonic energy to pass through.

(d) The tuned amplifier is less efficient.

(e) The tuned amplifier always introduces some distortion into the signal.

14. What type of circuit is shown in Fig. Test2-2? Assume the values of the resistors and capacitors are such that the circuit performs its intended function properly.

(a) A signal amplifier.

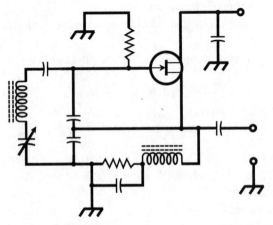

Fig. Test2-2. Illustration for Part Two Test Questions 14 through 18.

(b) A signal oscillator.

(c) A power supply.

(d) A wireless receiver.

(e) A waveform changer.

15. In the circuit of Fig. Test2-2, what should be connected to the terminal at the extreme upper right?

(a) A source of positive DC voltage.

(b) A source of AC voltage.

(c) The signal input.

(d) An oscillator.

(e) A rectifier.

16. In the circuit of Fig. Test2-2, what might be connected to the pair of terminals at the extreme lower right in a practical application?

(a) A power supply.

(b) A rectifier.

(c) A filter.

(d) A signal amplifier.

(e) Nothing. These terminals serve no purpose and might as well not be there.

17. In the circuit of Fig. Test2-2, what type of component is represented by the circle with the lines and arrow inside?

(a) An N-channel JFET.

(b) A P-channel JFET.

(c) An NPN bipolar transistor.

(d) A PNP bipolar transistor.

(e) A MOSFET.

18. If the variable capacitor in the circuit of Fig. Test2-2 is adjusted to increase its capacitance, what will happen?

(a) The waveform will become more like a sine wave.

(b) The circuit will amplify more.

(c) The circuit will produce more nearly pure DC.

(d) The frequency will decrease.

(e) Nothing. The variable capacitor serves no purpose and might as well not be there.

19. What form of memory can be accessed, but not overwritten, in the course of normal operation?

(a) Diode memory.

(b) JFET memory.

(c) Bipolar memory.

(d) Random-access memory.

(e) Read-only memory.

20. Generally, PNP and NPN circuits are similar except for
    (a) the signal frequencies and waveforms.
    (b) the type of battery and transformer required.
    (c) the method of coupling and the amplification factor.
    (d) the intensity of the oscillation and the stability of the frequency.
    (e) the polarities of the DC power-supply voltages and the directions of the resulting currents.

21. A Gunn diode would most likely be used in
    (a) a rectifier circuit.
    (b) a power-supply filter circuit.
    (c) an amplitude-limiting circuit.
    (d) a high-speed switch.
    (e) an oscillator circuit.

22. A quartz crystal would most likely be found in
    (a) a power supply.
    (b) an amplifier.
    (c) a rectifier.
    (d) a transformer.
    (e) an oscillator.

23. A PLL frequency synthesizer is known for its
    (a) high amplification factor.
    (b) pure DC output.
    (c) excellent voltage regulation.
    (d) sensitivity.
    (e) stability.

24. The middle electrode in a bipolar transistor is called
    (a) the base.
    (b) the gate.
    (c) the anode.
    (d) the cathode.
    (e) the tap.

25. What does Fig. Test2-3 most likely represent?
    (a) A characteristic curve for a resistor.
    (b) A characteristic curve for a diode.
    (c) A characteristic curve for a capacitor.

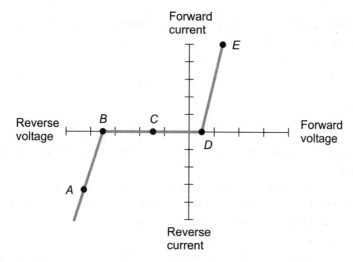

Forward
current

Reverse
voltage

Forward
voltage

Reverse
current

Fig. Test2-3.   Illustration for Part Two Test Questions 25 through 29.

(d) A DC waveform.
(e) An AC waveform.

26.   Which of the points in Fig. Test2-3 represents the forward breakover
point?
(a) *A*
(b) *B*
(c) *C*
(d) *D*
(e) *E*

27.   Which of the points in Fig. Test2-3 show a condition of reverse bias?
(a) *A*
(b) *B*
(c) *C*
(d) All three points *A*, *B*, and *C*.
(e) None of the three points *A*, *B*, or *C*.

28.   In the situation of Fig. Test2-3, at which point(s) does the device fail
to conduct?
(a) *A*
(b) *C*
(c) *E*
(d) The device does not conduct at any of the above three points.
(e) The device conducts at all of the above three points.

29. Which of the points in Fig. Test2-3 represents conduction as a result of avalanche effect?
   (a) *A*
   (b) *B*
   (c) *C*
   (d) *D*
   (e) *E*

30. The DC power input, in watts, to a bipolar-transistor amplifier circuit is equal to
   (a) the base current in amperes times the base voltage in volts.
   (b) the base current in amperes times the emitter voltage in volts.
   (c) the collector current in amperes times the collector voltage in volts.
   (d) the emitter current in amperes times the base voltage in volts.
   (e) Any of the above.

31. The electrodes in a JFET are called
   (a) the emitter, base, and collector.
   (b) the ground, tap, and anode.
   (c) the cathode, ground, and anode.
   (d) the positive, neutral, and negative terminals.
   (e) the source, gate, and drain.

32. A twin-T oscillator is commonly used to
   (a) produce high voltages.
   (b) generate audio signals.
   (c) adjust the gain of an amplifier.
   (d) receive weak signals.
   (e) convert AC to DC.

33. What type of circuit is shown in Fig. Test2-4? Assume the values of the resistors and capacitors are such that the circuit performs its intended function properly.
   (a) A signal amplifier.
   (b) A power supply.
   (c) A DC-to-AC converter.
   (d) A rectifier.
   (e) A filter.

34. What will happen if the polarities of the voltages shown in Fig. Test2-4 are reversed, that is, changed to −6 V instead of +6 V? Assume no other changes are made.

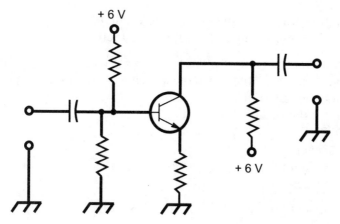

Fig. Test2-4.   Illustration for Part Two Test Questions 33 through 37.

(a) Nothing will happen; the circuit will continue working normally.
(b) The circuit will still work, but not as well.
(c) The circuit will stop working altogether.
(d) The circuit might work and it might not; it is anybody's guess.
(e) More information is needed to answer this question.

35. In Fig. Test2-4, what type of component is represented by the circle with the lines and arrow inside?
(a) An N-channel JFET.
(b) A P-channel JFET.
(c) An NPN bipolar transistor.
(d) A PNP bipolar transistor.
(e) A MOSFET.

36. What appears at, or should be connected to, the set of terminals on the right-hand side of the circuit shown in Fig. Test2-4?
(a) The input signal.
(b) The output signal.
(c) A high DC voltage.
(d) A low DC voltage.
(e) Nothing.

37. What does the capacitor on the left-hand side of the diagram in Fig. Test2-4 do?
(a) It blocks DC, but lets AC signals pass.
(b) It keeps the circuit from generating harmonics.
(c) It keeps the circuit from overheating.

(d) It ensures that the circuit will oscillate.

(e) Nothing. It might as well not be there.

38. Fill in the following sentence to make it true: "A _____ diode takes advantage of the avalanche effect to obtain power-supply voltage regulation."

(a) rectifier

(b) PNP

(c) Zener

(d) switching

(e) varactor

39. Amplification factor is commonly expressed in

(a) volts.

(b) amperes.

(c) watts.

(d) hertz.

(e) decibels.

40. One of the most serious shortcomings of MOSFETs is the fact that

(a) they are slow.

(b) they require extremely high current.

(c) they are easily destroyed by electrostatic discharge.

(d) they are inefficient.

(e) All of the above.

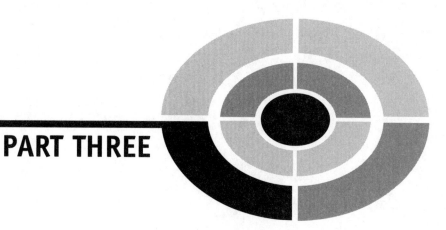

PART THREE

Wireless Electronics

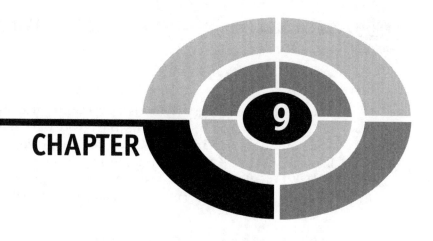

CHAPTER 9

Radio-Frequency Transmitters

A *radio-frequency (RF) transmitter* employs one or more oscillators to generate a radio-frequency (RF) signal, and amplifiers to generate the required power output. Some transmitters have signal mixers, too. In a transmitter, a mixer combines signals having two different frequencies to obtain output at a third frequency, which is either the sum or the difference of the input frequencies.

Modulation

Modulation is the process of "imprinting" or "impressing" data, sometimes called *intelligence*, onto an electric current or electromagnetic (EM) wave. The process can be done by varying the amplitude, frequency, or phase of the wave. Another method is to transmit a series of pulses, whose duration, amplitude, or spacing is made to vary. The way in which an EM wave is modulated is known as the *emission mode*, *emission type*, or simply *emission*.

THE CARRIER

The heart of a wireless signal is a *sine wave*, usually having a frequency far above the range of human hearing. This wave is known as the *carrier*. The lowest carrier frequency used for radio communications is a few kilohertz (kHz). The highest frequency is in the hundreds of gigahertz (GHz). For efficient data transfer, the carrier frequency must be at least 10 times the highest frequency of the modulating signal.

MORSE CODE

The simplest form of modulation is *on/off keying*. It is usually done in the oscillator of a *continuous-wave* (CW) radio transmitter. Figure 9-1 is a block diagram of a CW transmitter. This is the type of system most often used to transmit signals in Morse code. The use of "the code" or "Morse" in communications is still popular among some amateur radio operators.

Morse code is a *binary digital* mode. The duration of a Morse-code *dot* is equal to the duration of a *binary digit*, or *bit*. A *dash* is 3 bits long. The space between dots and dashes within a *character* is 1 bit. The standard space between characters in a *word* is 3 bits. The standard space between words is 7 bits. A punctuation symbol is sent as a character attached to the preceding word. An amplitude-versus-time rendition of the Morse word "eat" is shown in Fig. 9-2. The key-down (full-carrier) condition is called *mark*, and the key-up (no-signal) condition is called *space*.

Morse code is one of the slowest known methods of data transmission. Human operators use speeds ranging from about 5 words per minute (wpm) to 40 or 50 wpm. Machines, such as computers and data terminals, function at many times this rate. These systems usually employ *frequency-shift keying* (FSK) rather than on/off keying.

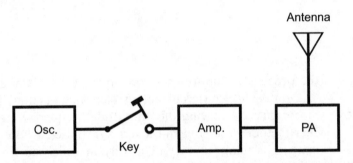

Fig. 9-1.   Block diagram of a CW Morse code transmitter.

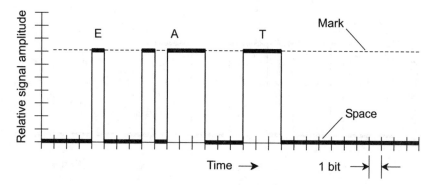

Fig. 9-2. The Morse code word "eat" as sent using CW.

RADIOTELETYPE

In FSK, the frequency of the signal is altered between two values nominally separated by a few hundred hertz. This separation is called the *shift*. In some systems, the carrier frequency is shifted between mark and space conditions. In other systems, a two-tone audio-frequency (AF) sine wave modulates the main carrier. The latter mode is called *audio-frequency-shift keying* (AFSK).

There are two common codes commonly used with FSK and AFSK: *Baudot* (pronounced "baw-DOE") and *ASCII* (pronounced "ASK-ee"). The acronym ASCII stands for *American Standard Code for Information Interchange*. The main difference between these two codes is that ASCII allows for more symbols and characters.

In *radioteletype* (RTTY) FSK and AFSK systems, a *terminal unit* (TU), also known as a *modem* (a contraction of the technical term *modulator/demodulator*), converts received signals into electrical impulses that operate a teleprinter, or that display characters on a computer screen. The TU also generates the signals necessary to send RTTY as an operator types on a keyboard. A frequency-versus-time graph of the word "eat," sent using FSK with Morse code, is shown in Fig. 9-3. A block diagram of an AFSK transmitter is shown in Fig. 9-4.

AMPLITUDE MODULATION

A voice signal is a complex waveform with frequencies mostly in the range between 300 Hz and 3000 Hz. The instantaneous amplitude of a carrier can be varied, or modulated, by these waveforms, thereby transmitting voice information. Figure 9-5 shows a bipolar-transistor circuit for obtaining

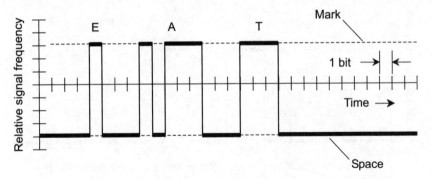

Fig. 9-3.   The Morse code word "eat" as sent using FSK.

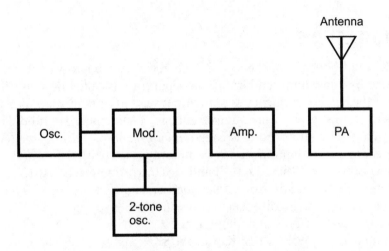

Fig. 9-4.   Block diagram of an AFSK transmitter.

amplitude modulation (AM). This circuit works efficiently provided the AF input signal is not too strong. If the AF input is too strong, distortion occurs.

Two complete AM transmitters are shown in block-diagram form in Fig. 9-6. At A, *low-level AM* is illustrated. In this mode, all the amplifier stages following the modulator must be linear, so they do not distort the modulating waveform. In some transmitters, AM is done in the last power amplifier (PA) stage, as shown in Fig. 9-6B. The PA operates in class C, and plays a dual role: modulator and final amplifier. Because there are no amplification stages beyond the modulator in this type of system, there is no need to worry about linearity in the amplifiers. This scheme is *high-level AM*. Despite the fact that it's been around for a long time, it's still used in some broadcast stations because it works well!

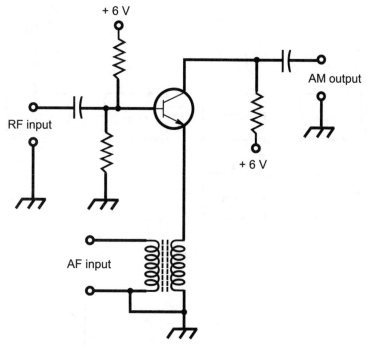

Fig. 9-5.   An amplitude modulator circuit.

MODULATION PERCENTAGE AND SIDEBANDS

The extent of AM is expressed as a percentage, from 0% (an unmodulated carrier) to 100% (full modulation). In an AM signal modulated 100%, 1/3 of the power is used to convey the data. The other 2/3 is consumed by the carrier wave, which does not contribute to the transmitted intelligence. Increasing the modulation past 100% causes distortion of the signal, degrades the efficiency of the transmitter, and causes the signal to be spread out over an unnecessarily wide band of frequencies.

In Fig. 9-7, a *spectral display* for an AM voice radio signal is illustrated. The horizontal scale is calibrated in increments of 1 kHz per division. The vertical scale is relative. The AF components, containing the intelligence transmitted, appear as *sidebands* on either side of the carrier. The RF energy that occurs between −3 kHz and the carrier frequency constitutes the *lower sideband (LSB)*. The RF from the carrier frequency to +3 kHz represents the *upper sideband (USB)*. The *bandwidth* of the RF signal is the difference between the maximum and minimum sideband frequencies.

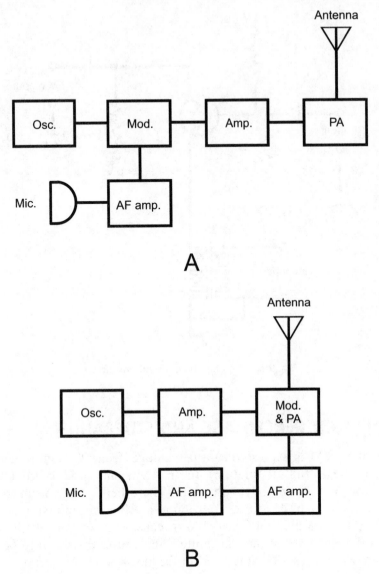

Fig. 9-6.   At A, low-level AM. At B, high-level AM.

In an AM signal, the bandwidth is twice the highest audio modulating frequency. In the example of Fig. 9-7, the voice energy is at or below 3 kHz; thus the bandwidth of the complete signal is 6 kHz. This is typical of an analog voice communications signal. In AM broadcasting in which music is transmitted, the energy exists over a wider bandwidth. The same is true of high-speed data.

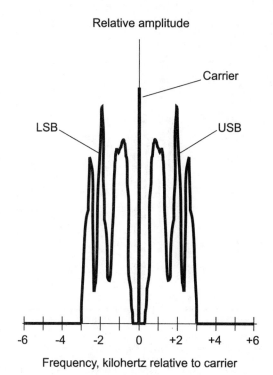

Fig. 9-7.   Spectral display of a typical AM voice signal.

SINGLE SIDEBAND

In AM, most of the power is used up by the carrier. The two sidebands are mirror-image duplicates. If the carrier and one of the sidebands is eliminated, all of the available power goes into data transmission, and the bandwidth can be reduced by 50% or more. The remaining voice signal has a spectral display resembling Fig. 9-8. This is *single-sideband* (SSB) transmission. Either the LSB or the USB can be used; either mode works as well as the other. The example in Fig. 9-8 shows LSB emission.

An SSB transmitter employs a *balanced modulator*. This circuit works like an amplitude modulator, except the carrier is phased out. A balanced modulator circuit is shown in Fig. 9-9. The RF carrier input signal is split and applied to the bases of the two transistors, 180° out of phase with each other. They are out of phase because they are taken from opposite ends of the input transformer. This causes the carrier waves from the collectors of the two transistors, which are directly connected, to cancel each other. The cancellation does not occur with respect to the AF input signal, or with the part of the RF signal attributable to it. This leaves only sideband energy,

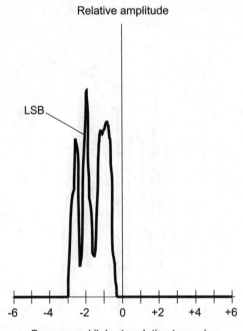

Relative amplitude

LSB

-6   -4   -2   0   +2   +4   +6

Frequency, kilohertz relative to carrier

Fig. 9-8.   Spectral display of a voice LSB signal.

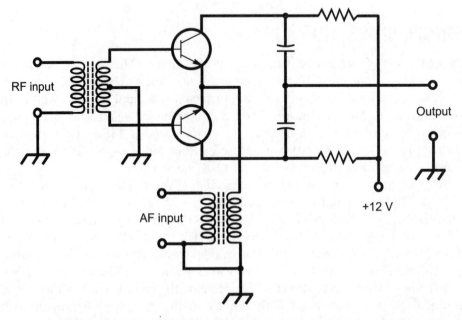

Fig. 9-9.   A balanced modulator circuit using bipolar transistors.

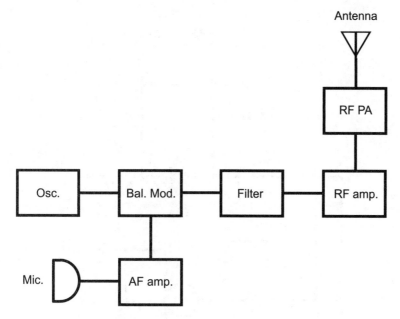

Fig. 9-10. Block diagram of an SSB transmitter.

and no carrier energy, in the output signal. One of the sidebands is removed by a *bandpass filter* in a later stage of the transmitter. Figure 9-10 is a block diagram of a basic SSB transmitter, including the bandpass filter and PA.

High-level modulation cannot be used for SSB transmission. The balanced modulator is always placed in a low-power section of the transmitter. The RF amplifiers that follow any amplitude modulator must all be linear to avoid distortion. The RF amplifiers following a balanced modulator generally work in class A, except for the PA, which functions in class AB or class B.

FREQUENCY AND PHASE MODULATION

In *frequency modulation* (FM), the overall amplitude of the signal remains constant, and the instantaneous frequency is varied. Class-C power amplifiers can be used in FM transmitters without introducing distortion, because the amplitude does not fluctuate. Thus, linearity is of no concern.

The most direct way to get FM is to apply the audio signal to a *varactor* in a tuned oscillator. An example of this scheme, known as *reactance modulation*, is shown in Fig. 9-11. The varying voltage across the varactor

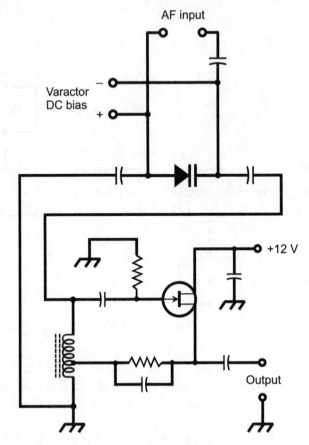

Fig. 9-11.   Reactance modulation to obtain FM.

causes its capacitance to change in accordance with the audio waveform. The changing capacitance results in variation of the resonant frequency of the inductance-capacitance (LC) tuned circuit, causing a swing in the frequency generated by the oscillator. In this example, the oscillator is a Hartley circuit that uses an N-channel JFET.

An indirect way to get FM is to modulate the phase of the oscillator signal. Any change in the instantaneous phase of a sine-wave RF carrier causes a change in its instantaneous frequency. (Think of a change in phase as being analogous to a change in position, and a change in frequency as being analogous to a change in speed. You can't have one without some of the other! But frequency and phase modulation are not exactly the same thing, just as position and motion are not identical.)

When *phase modulation* is used, the audio signal must be processed, adjusting the amplitude-versus-frequency response of the audio amplifiers.

Otherwise the signal sounds unnatural when it is heard at the output of an FM receiver.

DEVIATION

Deviation is the maximum extent to which the instantaneous carrier frequency differs from the unmodulated-carrier frequency. For most FM voice transmitters, the deviation is standardized at ±5 kHz (Fig. 9-12). The deviation obtainable by means of direct FM is greater, for a given oscillator frequency, than the deviation that can be obtained using phase modulation. Deviation can be increased by a *frequency multiplier*. When an FM signal is passed through a frequency multiplier, the deviation is multiplied along with the carrier frequency.

In FM hi-fi music broadcasting, and in some other applications, the deviation is much greater than ±5 kHz. This is called *wideband FM*, as opposed to *narrowband FM* discussed above. The deviation for an FM signal should be at least equal to the highest modulating audio frequency if optimum fidelity is to be obtained. Thus, ±5 kHz is more than enough for voice. For music, a deviation of at least ±15 kHz to ±20 kHz is needed.

The ratio of the frequency deviation to the highest modulating audio frequency is called the *modulation index*. Ideally, this figure should be between 1:1 and 2:1. If it is less than 1:1, the signal sounds muffled, and

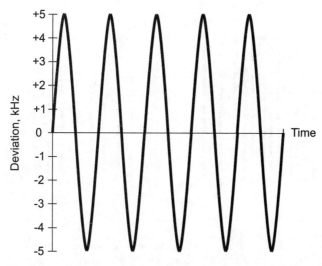

Fig. 9-12.   Frequency-versus-time rendition of an FM signal.

efficiency is sacrificed. Increasing it beyond 2:1 broadens the bandwidth without providing much improvement in the signal quality.

PULSE MODULATION

Several types of *pulse modulation* (PM) signals are shown in Fig. 9-13 as amplitude-versus-time graphs. In PM, bursts of carrier are transmitted. Instead of modulating the carrier wave itself, some characteristic of the pulses is varied, such as the strength of each pulse, the duration of each pulse, or the time interval between consecutive pulses.

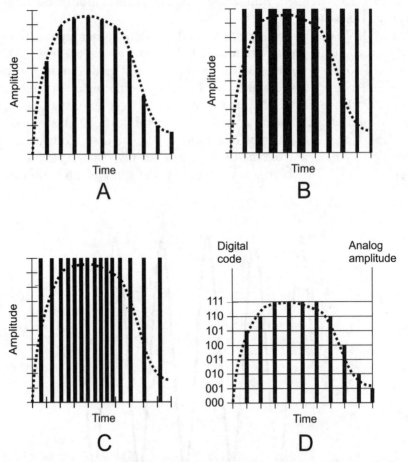

Fig. 9-13. At A, pulse amplitude modulation; at B, pulse width modulation; at C, pulse interval modulation; at D, pulse code modulation. Dashed curves represent the analog input. Vertical bars represent pulses.

In *pulse amplitude modulation* (PAM), the strength of each individual pulse varies according to the modulating waveform (Fig. 9-13A), and the pulses all have equal duration. *Pulse duration modulation* (PDM), also called *pulse width modulation* (PWM), is shown at B. All the pulses in this mode have equal amplitude; the energy in each pulse varies because some pulses last longer than others. *Pulse interval modulation* (PIM), also called *pulse frequency modulation* (PFM), is shown at C. In this mode, the amplitude and duration of every pulse is identical, but the signal energy varies because the pulses occur more or less often. The pulses are generated most often at modulation peaks (maxima), and least often at modulation troughs (minima).

In *pulse-code modulation* (PCM), any of the aforementioned aspects – amplitude, duration, or frequency – of a pulse train can be varied. But rather than having a theoretically infinite number of possible states or levels, there are only a few. The PCM mode is digital, rather than analog. Usually, the number of possible levels in PCM is some power of 2, such as $2^3 = 8$ or $2^4 = 16$. Figure 9-13D shows an example of 8-level PCM in which the pulse amplitude is varied.

PROBLEM 9-1

Suppose an FM signal is generated by means of phase modulation, in which the effective deviation is ± 2.5 kHz. How can this be increased to the standard ± 5 kHz for voice transmission?

SOLUTION 9-1

The signal can be passed through a *frequency doubler*, which is a circuit that produces an output signal having twice the frequency of the input signal. When this is done, the deviation is doubled along with the signal frequency. The input signal varies above and below the unmodulated carrier frequency by 2.5 kHz, and the output signal varies above and below the unmodulated carrier frequency by (2.5×2) kHz, or 5.0 kHz.

PROBLEM 9-2

What happens to an FSK signal when it is passed through a frequency doubler?

SOLUTION 9-2

The shift, along with the carrier frequency, is doubled. Thus, for example, if the input signal has a shift of 85 Hz between the mark and space conditions, the output signal has a shift of (85×2) Hz, or 170 Hz.

Analog-to-Digital Conversion

Pulse-code modulation, such as is shown at Fig. 9-13D, is a form of *analog-to-digital (A/D) conversion*. A voice signal, or any continuously variable signal, can be digitized, or converted into a string of pulses, whose amplitudes can achieve only a finite number of states. A circuit that performs A/D conversion is called an *A/D converter* or *ADC*.

SAMPLING RESOLUTION

In A/D conversion, the number of states is always a power of 2. Thus, each state can be represented as a binary-number code. Fidelity improves as the number of possible states increases. The number of states is called the *sampling resolution*, or simply the *resolution*. A resolution of $2^3 = 8$ (as shown in Fig. 9-13D) is good enough for voice transmission, and is the standard for commercial digital voice circuits. A resolution of $2^4 = 16$ is adequate for high-fidelity (hi-fi) music reproduction.

SAMPLING RATE

The fidelity of a digital signal also depends on the frequency at which sampling is done. In general, the *sampling frequency*, more often called the *sampling rate*, must be at least twice the highest data frequency. For an audio voice signal with components as high as 3 kHz, the minimum sampling rate for effective digitization is 6 kHz; the commercial voice standard is 8 kHz. For hi-fi digital transmission, the standard sampling rate is 44.1 kHz, which is a little more than twice the highest audible frequency of 20 kHz.

PROBLEM 9-3
What does a digitized voice signal sound like if the sampling resolution is insufficient, or if not all of the pulses are received all of the time?

SOLUTION 9-3
The signal may sound fluttery, as if the other person is talking to you through a long, hollow tube, or as if wind is blowing against the microphone. This phenomenon is often observed with cell-phone signal reception when

the connection is poor, because digital pulses are occasionally dropped during rapid signal fades.

Image Transmission

Non-moving images such as photographs and drawings can be sent within the same bandwidth as voice signals. For detailed images, or for full-motion video, the necessary bandwidth is greater.

FACSIMILE

Non-moving images are transmitted by *facsimile*, also called *fax*. If data is sent slowly enough, any amount of detail can be transmitted within a 2700-Hz-wide band. This is how telephone fax works, for example. In general, the greater the detail in a fax image, for a constant bandwidth such as 2700 Hz, the longer it takes to send the fax.

To send an image by fax, a hard copy document or photo is wrapped around a *drum*. The drum is rotated at a slow, controlled rate. A spot of light scans from left to right; the drum moves the document so a single line is scanned with each pass of the light spot. This continues, line by line, until the complete *frame*, or picture, has been scanned. The reflected light is picked up by a *photodetector*. Dark parts of the image reflect less light than bright parts, so the current through the photodetector varies. This current modulates a carrier.

Alternatively, a fax can be sent by scanning the document with a computer scanner, and using the computer to process the image into a form suitable for transmission over the telephone line.

ANALOG FAST-SCAN TELEVISION

To get a realistic impression of motion, it is necessary to transmit at least 20 complete *frames* (stationary images) per second, and the detail must be adequate. A conventional *fast-scan television* (FSTV) system usually provides 30 frames per second. There are 525 or 625 horizontal *scanning lines* in each frame. In *high-definition television* (HDTV), there are more lines; in *slow-scan television* (SSTV) there are fewer lines. The images traditionally have a horizontal-to-vertical length ratio, or *aspect ratio*,

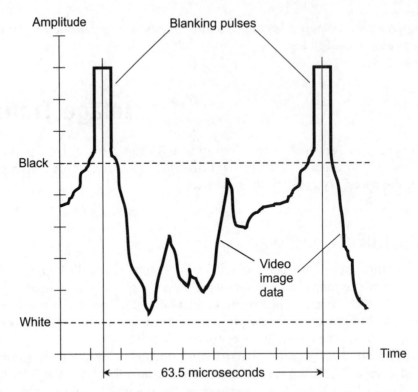

Figure 9-14.   A conventional analog FSTV signal, showing one line of data.

of 4 : 3, although in some newer systems this ratio is larger. Each line contains shades of brightness in a gray-scale system, and shades of brightness and hue in a color system. In FSTV broadcasting, the image is sent as an AM signal, and the sound is sent as an FM signal. In SSTV communications, SSB is the most often-used mode.

Because of the large amount of information sent, an FSTV channel is wide. A standard FSTV channel in the North American system takes up 6 MHz of spectrum space. That's over 2000 times the bandwidth of an SSB voice signal! Figure 9-14 shows a typical line of data in an FSTV signal as it would appear on an oscilloscope.

An FSTV transmitter consists of a camera, an oscillator, an ampli-tude modulator, and a series of amplifiers for the video signal. The audio system consists of an input device (such as a microphone), an oscillator, a frequency modulator, and a feed system that couples the RF output into the video amplifier chain. There is also an antenna or cable output. Figure 9-15 is a block diagram of an analog FSTV transmitter.

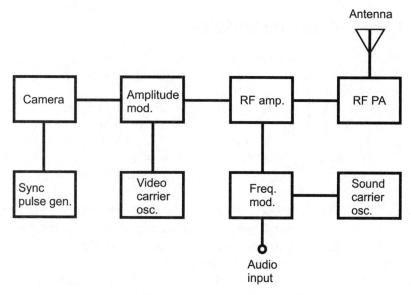

Fig. 9-15.   Block diagram of a conventional analog FSTV transmitter.

ANALOG SLOW-SCAN TELEVISION

It is possible to send a video image in a band much narrower than 6 MHz. In SSTV, this is done by greatly reducing the rate at which the frames are transmitted. In addition, the image resolution (detail) is reduced. An SSTV signal is sent within about 2700 Hz of spectrum, the same as needed by an SSB voice signal or fax signal. An SSTV signal typically contains one frame every 8 seconds. There are 120 lines per frame.

The modulation for black-and-white, more properly called *grayscale*, SSTV in radio communications is obtained by inputting audio signals into an SSB transmitter. An AF sine wave having a frequency of 1500 Hz corresponds to black; a sine wave having a frequency of 2300 Hz corresponds to white. Intermediate audio frequencies produce shades of gray. Synchronization signals are sent at 1200 Hz. These are short bursts, lasting 0.030 seconds (30 milliseconds) for vertical synchronization and 0.005 seconds (5 milliseconds) for horizontal synchronization.

Sometimes, SSTV images are sent along with voice data. The video is sent on one sideband, while the audio is sent on the other sideband.

The differences between color SSTV and grayscale SSTV are similar to the differences between color FSTV and grayscale FSTV. Separate signals are sent for the red (R), green (G), and blue (B) primary colors.

HIGH-DEFINITION TELEVISION

The term *high-definition television* (HDTV) refers to any of several similar methods for getting more detail into a TV picture, and for obtaining better audio quality, compared with standard FSTV.

A standard FSTV picture has 525 lines per frame, but HDTV systems have many more. The image is scanned more often, too. High-definition TV is a digital mode; this offers another advantage over conventional FSTV. Digital signals propagate better, are easier to deal with when they are weak, and can be processed in ways that analog signals cannot.

Some HDTV systems use *interlacing* in which two *rasters*, or sets of scanning lines, are "meshed" together. This effectively doubles the image resolution without doubling the cost of the hardware. But this scheme can cause annoying *jitter* in fast-moving pictures.

New ideas are constantly developing in HDTV technology. In the year 2003, Japanese engineers first demonstrated a large-screen system called *ultra high-definition video* (UHDV). Other advanced, full-motion video technologies render three-dimensional (3D) images. The main limitations on cutting-edge video technologies are the bandwidth, which increases with definition and with 3D imagery, and the cost, which is beyond the reach of most private citizens when the first prototypes are built, tested, and demonstrated.

DIGITAL SATELLITE TV

In recent years, a new form of television, called *digital satellite TV*, has taken hold throughout much of the world. We'll look at this briefly in the next chapter when we discuss how RF signals are received.

PROBLEM 9-4
How might the bandwidth of a conventional analog FSTV video signal be cut roughly in half without sacrificing any of the image detail, and without slowing down the scanning speed?

SOLUTION 9-4
A conventional analog FSTV video signal is transmitted using AM. This means there is a carrier wave, along with the LSB and USB. If the signal were converted to SSB, its bandwidth could be reduced, in theory, from 6 MHz to 3 MHz.

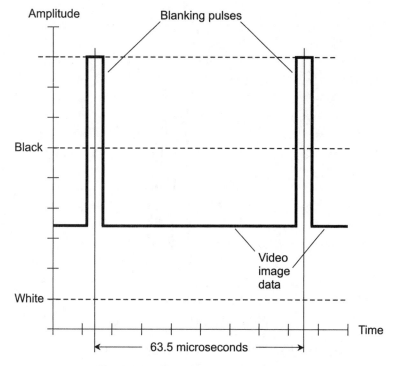

Fig. 9-16.   Illustration for Problem 9-5.

PROBLEM 9-5

Suppose an FSTV camera is pointed at a featureless, gray, overcast sky or a blank gray wall. What does a time-domain display of the video signal, such as is displayed on an oscilloscope, look like in this situation?

SOLUTION 9-5

The video signal amplitude between blanking pulses is constant, somewhere between white and black. The video signal, when observed on an oscilloscope, therefore looks something like the illustration of Fig. 9-16.

Quiz

Refer to the text in this chapter if necessary. Answers are in the back of the book.

1. What does Fig. 9-17 represent?

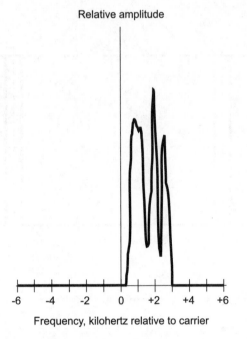

Relative amplitude

Frequency, kilohertz relative to carrier

Fig. 9-17.   Illustration for Quiz Questions 1 and 2.

    (a) An oscilloscope display of an FM voice signal.
    (b) A time-domain display of an FSTV video signal.
    (c) A spectral display of a USB voice signal.
    (d) A digital display of a CW Morse code signal.

2.   What is the approximate bandwidth of the signal shown in Fig. 9-17?
    (a) 2.7 Hz
    (b) 270 Hz
    (c) 2700 Hz
    (d) It is impossible to tell from the information shown.

3.   For good fidelity at the receiving end, the deviation for an FM signal should be
    (a) less than half of the highest modulating audio frequency.
    (b) at least equal to the highest modulating audio frequency.
    (c) less than half of the lowest modulating audio frequency.
    (d) zero, because there should be no carrier output.

4.   The detail that can be sent in a fax signal whose bandwidth is 2700 Hz is
    (a) limited by the time available to send and receive the fax.

(b) limited by the emission type.

(c) limited by the complexity of the image.

(d) limited to extremely low image resolution.

5. The maximum extent to which the instantaneous carrier frequency differs from the unmodulated-carrier frequency in an FM signal is called the

(a) deviation.

(b) bandwidth.

(c) modulation.

(d) sideband.

6. Morse code is usually transmitted using

(a) CW emission.

(b) AM emission.

(c) SSB emission.

(d) FSTV emission.

7. When a balanced modulator is used to generate an SSB signal, the output carrier wave takes up

(a) essentially none of the power.

(b) approximately 1/3 of the output power.

(c) approximately 2/3 of the power.

(d) essentially all of the power.

8. A raster is a set of

(a) tones used in AFSK.

(b) sidebands used in SSB.

(c) scanning lines in a TV frame.

(d) frequencies used in SSTV.

9. An oscilloscope normally has

(a) a time-domain display.

(b) a bandwidth of 2700 Hz.

(c) a bandwidth of 6 MHz.

(d) a bandwidth of half that of a typical AM signal.

10. Which of the following is most often transmitted in a binary (two-level) digital mode?

(a) USB emission.

(b) LSB emission.

(c) FSTV emission.

(d) CW emission.

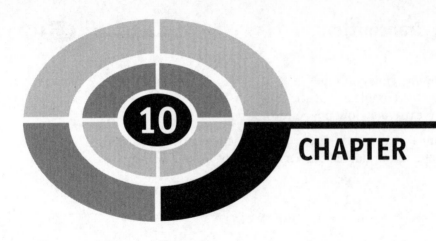

CHAPTER

Radio-Frequency Receivers

A *radio-frequency (RF) receiver* converts EM waves into the original messages sent by a distant RF transmitter. In the broadest sense, a receiver "undoes" what a transmitter does.

Simple Designs

Some amazingly simple circuits can work as EM signal receivers (and some receivers can be incredibly complicated). Here are three examples of basic receiver schemes.

CRYSTAL SET

When an RF diode is connected in a circuit such as the one shown in Fig. 10-1, the result is a device capable of picking up strong AM signals.

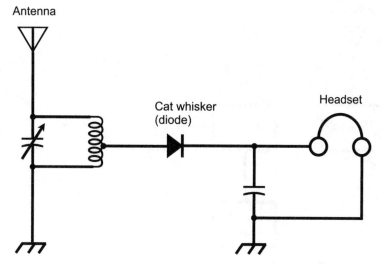

Fig. 10-1.   A crystal-set radio receiver requires no source of power other than the incoming signal.

The diode, sometimes called a *crystal*, has given rise to the nickname *crystal set*. There is no source of power except the incoming signal, which comes from the antenna. The output is sufficient to drive an earphone or headset if the transmitting station is within a few kilometers of the receiver, and if the antenna is large.

The diode acts as a *detector*, also called a *demodulator*, to recover the modulating waveform from the signal. If the detector is to be effective, the diode must have low capacitance, so that it works as a rectifier at RF but not as a capacitor.

DIRECT-CONVERSION RECEIVER

A *direct-conversion receiver* derives its output by mixing incoming signals with the output of a variable-frequency *local oscillator* (LO). The received signal is fed into a *mixer*, along with the output of the LO. The mixer also serves as the detector. Figure 10-2 is a simplified block diagram of a direct-conversion receiver.

For reception of CW Morse code signals, the LO is set slightly above or below the signal frequency. The incoming signal and the LO output mix, producing a signal at a *beat frequency* equal to the difference between the LO and signal frequencies. For reception of SSB signals, the LO is set to exactly the incoming signal frequency, called the *carrier frequency*. This exact

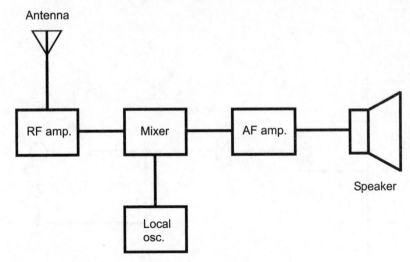

Fig. 10-2. Block diagram of a direct-conversion radio receiver.

matching of frequencies is called *zero beat*, because the beat frequency is equal to 0.

A direct-conversion receiver is not very good at differentiating between two signals whose signals are close together. That is to say, the *selectivity* is poor. This is because signals on either side of the LO frequency interfere with one another.

SUPERHETERODYNE RECEIVER

A *superheterodyne receiver* uses one or more local oscillators and mixers to obtain an output signal that is always at the same frequency, or that is within a fixed, narrow range (or *band*) of frequencies. When this is done, all incoming signals can be processed using circuits optimized to work within that constant, narrow band.

In the superheterodyne receiver or *superhet*, the incoming signal from the antenna is first passed through a tunable, sensitive amplifier called the *front end*. The output of the front end is mixed with the signal from a tunable LO. Either the sum or the difference signal is amplified. This signal is at the *first intermediate frequency* (or *first IF*), and can be filtered to obtain a high degree of selectivity.

If the first IF signal is detected, the radio is known as a *single-conversion superhet* because mixing takes place one time in the whole system. Some receivers use a second mixer and second LO, converting the first IF signal to a lower-frequency signal at the *second IF*. This type of system is called a *double-conversion superhet*.

In any superhet, a sophisticated *IF bandpass filter* is designed for use on a fixed frequency, resulting in excellent selectivity. Some bandpass filters allow for adjustable bandwidth. The sensitivity of the superhet is excellent as long as the fixed-frequency IF amplifiers are all tuned to the same frequency. The process of tuning all the IF stages to work together, and the maintenance of this optimal condition, is called *receiver alignment*.

A superheterodyne receiver can sometimes intercept or generate unwanted signals. False signals that originate outside the receiver are called *images*. Internally generated, false signals are called *birdies*. If the LO frequencies are judiciously chosen, images and birdies are rarely a problem.

PROBLEM 10-1
Refer to the crystal set radio receiver diagram (Fig. 10-1). What will happen if lightning strikes somewhere near the antenna, causing a current surge that damages the diode so it no longer acts as a rectifier at RF?

SOLUTION 10-1
The diode will no longer be able to separate the audio information from the incoming signal. Thus, no AF energy will appear in the headphones, and the listener will hear nothing.

The Modern Receiver

Wireless communications receivers operate over specific ranges, known as bands, of the radio spectrum. The particular set of bands that a receiver covers depends on the application for which the receiver is designed.

SPECIFICATIONS

The *specifications* of a receiver indicate how well it can do the things it's designed to do. Here are some specifications that are often quoted for radio receivers.

- *Sensitivity*: The most common way to express receiver sensitivity is to state the number of microvolts (millionths of a volt, symbolized μV) that must exist at the antenna terminals to produce a certain *signal-to-noise ratio* (S/N) or *signal-plus-noise-to-noise ratio* (S+N/N) for the listener. Sensitivity is related to the gain of the front end, but the amount of noise the front-end transistor generates is more significant, because subsequent stages amplify the front-end noise output as well as the signal output.

- *Selectivity*: The width of a receiver *passband* (the range of signal frequencies that is allowed through the system at any given time) is established by a wideband *preselector* in the early RF amplification stages, and is narrowed down by filters in later amplifier stages. The preselector makes the receiver sensitive within a range of plus-or-minus ($\pm$) a few percent of the desired signal frequency. The narrowband filter passes only the desired signal; signals in nearby channels are rejected.

- *Dynamic range*: This is a measure of the extent to which a receiver can maintain a fairly constant output, and nevertheless keep its rated sensitivity, in the presence of signals ranging from weak to strong. In a receiver with poor dynamic range, weak signals come in all right as long as there are no strong signals nearby in frequency, but a strong signal overwhelms the circuits, and the weak signal is obliterated.

- *Noise figure*: This is a measure of the amount of noise a radio generates inside its own circuits (as opposed to noise coming from the environment). The less internal noise a receiver produces, in general, the lower is the noise figure, and the better is the S/N ratio. This specification becomes more and more important as the received frequency goes up. At very high frequencies (VHF) and above – that is, over about 30 MHz – noise figure is of paramount importance.

OVERVIEW OF THE SINGLE-CONVERSION SYSTEM

Figure 10-3 is a block diagram of a single-conversion superheterodyne radio receiver. Individual designs vary, but this circuit is representative.

- *Front end*: This consists of the first RF amplifier, and usually includes selective filters between the amplifier and the antenna. The dynamic range and sensitivity of a receiver are largely determined by the performance of the front end.

- *Mixer*: This circuit converts the variable signal frequency to a constant IF. The output is either the sum or the difference of the signal frequency and the LO frequency.

- *IF stages*: Here's where most of the amplification takes place. These stages are also where sharp selectivity can be obtained. Specialized filters are used to get the desired bandwidth and response. In more advanced receivers, the sharp selectivity is the result of *digital signal processing (DSP)*.

- *Detector*: The detector extracts the information, also called intelligence, from the signal. The output of an ideal detector is identical to the modulating signal that comes from the transmitting station.
- *Post-detector stages*: Following the detector, one or two stages of audio amplification boost the signal to a volume suitable for listening with a speaker or headset. Alternatively, the signal can be fed to a printer, fax machine, video display, or computer.

PROBLEM 10-2

Imagine a superheterodyne receiver built according to the design shown in Fig. 10-3. What will happen if the local oscillator frequency becomes unstable?

SOLUTION 10-2

Assuming the intended received signal has a constant, stable frequency, any change in the LO frequency will produce a similar change in the difference frequency at the output of the mixer. This will make the received signal frequency fluctuate in the IF stages. As a result, reception will be difficult or impossible.

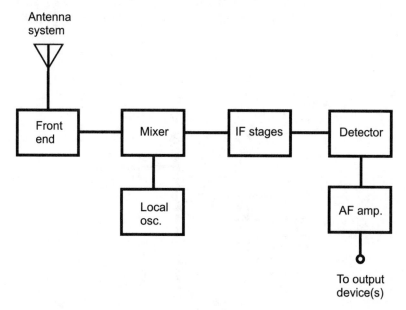

Fig. 10-3.   Block diagram of a single-conversion superhet.

Pre-detector Stages

In any RF receiver, the stages preceding the detector must be designed so they provide reasonable gain, but they should produce as little noise as possible.

PREAMPLIFIER

All preamplifiers operate in class A, and most employ FETs. Figure 10-4 shows a simple RF preamplifier. Input tuning provides some selectivity. This circuit produces 5 dB to 10 dB gain, depending on the frequency and the choice of FET.

It is important that a receiving preamplifier be linear. Nonlinearity results in *intermodulation distortion* (IMD) that can produce false signals, raise the background noise level, and reduce the dynamic range.

FRONT END

At low and medium frequencies, there is considerable atmospheric noise caused by lightning and other environmental disturbances. Above 30 MHz, atmospheric noise diminishes, and the main factor that limits receiver sensitivity is noise generated within the circuits. For this reason, front-end design becomes increasingly critical as the frequency rises, especially above 30 MHz.

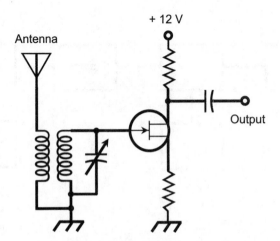

Fig. 10-4.   Schematic diagram of an RF preamplifier.

The front end, like a preamplifier, must be as linear as possible; the greater the degree of nonlinearity, the more susceptible the circuit is to the generation of IMD. The front end should also have the greatest possible dynamic range. If the dynamic range is not sufficient, a strong local signal can cause the receiver to *desensitize* (lose sensitivity) over much or all of its operating frequency range.

PRESELECTOR

The preselector provides a bandpass response, or degree of selectivity, that improves the S/N ratio and reduces the likelihood of receiver desensitization by a strong signal far removed from the operating frequency. In a superheterodyne receiver, the preselector also helps the system reject image signals.

A preselector can be tuned by means of *tracking* with the tuning dial, so it is always adjusted correctly as the receiver is tuned from one frequency to another. This requires careful design and alignment, and can be difficult to achieve. Some receivers incorporate preselectors that must be adjusted independently of the main tuning control. The alignment of this type of receiver is less critical, but the preselector must be "tweaked" every time a significant change is made in the frequency.

PROBLEM 10-3
Suppose a receiver has a preselector that tracks along with the main tuning. What will happen if such a receiver gets out of alignment, so the passband of the preselector does not always coincide with the frequency of the desired incoming signal?

SOLUTION 10-3
When this happens, the overall gain of the receiver will be degraded at some frequencies. If the misalignment is bad enough, the sensitivity and dynamic range may also suffer.

MIXERS

A mixer requires a waveform-distorting circuit element of some kind. This makes it easy for the signals to beat against each other and produce energy at the sum and difference frequencies. These signals are known as *mixing products*.

When non-amplifying elements such as diodes are used to mix signals, the circuit is called a *passive mixer*. It does not require an external source of power, and there is some loss when it is inserted into a system. An *active mixer* employs one or more transistors or integrated circuits to obtain some gain as well as to mix the signals. The amplifying device is usually biased in class AB or B.

The output of the mixer circuit can be tuned to either the sum frequency or the difference frequency, as desired.

IF STAGES

A high IF (several megahertz) is preferable to a low IF (less than 1 MHz) to suppress unwanted image signals. This suppression is called *image rejection*. But a low IF is better for obtaining sharp selectivity. A double-conversion receiver has a comparatively high first IF and a low second IF to get the "best of both worlds."

Several IF amplifiers can be *cascaded*, or connected one after another, to obtain high gain and high selectivity. Transformer coupling is used between stages. These amplifiers come after the mixer stage and before the detector stage. Double-conversion receivers have two sets, called *chains*, of IF amplifiers. The *first IF chain* follows the first mixer and precedes the second mixer. The *second IF chain* follows the second mixer and precedes the detector.

The selectivity of an IF chain can be quantified. The bandwidths are compared for two power-attenuation values of an incoming signal, usually 3 dB down and 30 dB down with respect to the gain obtained when the signal is at the center of the passband. The ratio of the 30 dB selectivity to the 3 dB selectivity is called the *shape factor*. The smaller the shape factor, the more rapidly the gain decreases as a signal moves from within the passband to outside it. A small shape factor, such as 2 : 1 or less, is good because it produces the best rejection of unwanted signals while allowing all of the desired signal to pass through.

Detectors

Detection, also called *demodulation*, is the recovery of intelligence such as audio, images, or printed data from a signal.

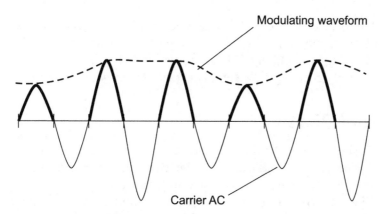

Fig. 10-5A. A simplified drawing of how envelope detection takes place.

DETECTION OF AM

The modulating waveform can be extracted from an AM signal by rectifying it. A simplistic view of this is shown in Fig. 10-5A. The rapid pulsations occur at the carrier frequency. The slower fluctuation (dashed line) is a duplication of the modulating intelligence. The carrier pulsations are smoothed out by passing the output through a capacitor large enough to hold the charge for one carrier current cycle, but not so large that it smooths out the cycles of the modulating signal. This scheme is known as *envelope detection*.

Here's a more technical description of what happens in AM demodulation. Any AM signal contains a carrier, which is an RF sine wave, and sidebands, which are signals that appear at frequencies slightly above and below that of the carrier. The sidebands are produced in the transmitted AM signal when the carrier beats against intelligence, such as voices, music, or encoded video imagery. An AM envelope detector separates the energy contained in the sidebands from the carrier, which itself contains energy but no information. The carrier is, in effect, thrown away. The resulting output is a replica of the original intelligence.

DETECTION OF CW

For detection of CW Morse code signals, it is necessary to inject a signal into the receiver a few hundred hertz from the carrier. The injected signal is produced by a tunable *beat-frequency oscillator* (BFO). The BFO signal and the desired CW signal are mixed to produce audio output at the difference frequency.

In order to receive a keyed Morse-code CW signal, the BFO should be tuned to a frequency that results in a comfortable listening pitch. For most people this is approximately 700 Hz above or below the carrier frequency. This method of receiving CW signals is called *heterodyne detection*.

DETECTION OF FSK

Frequency-shift keying (FSK) can be detected using the same method as CW detection. The carrier beats against the BFO in the mixer, producing an audio tone that alternates between two different pitches.

With FSK, the BFO frequency is set a few hundred hertz above or below both the mark and space carrier frequencies. The *frequency offset*, or difference between the BFO and signal frequencies, determines the audio output frequencies, and must be set so standard tone pitches result.

Unlike the situation with CW reception, there is little tolerance for variation in the BFO adjustment when receiving an FSK signal. The BFO must be set at exactly the correct frequency, so the audio output tones occur at the correct pitches. Otherwise the modem will be "deaf" to the detector output.

DETECTION OF FM

Frequency-modulated (FM) signals can be detected in various ways. Some schemes work better than others, but in general, the more effective methods also require the most complex and expensive circuits. Here are four common systems for detection of FM.

- *Slope detection*: An AM receiver can be used to detect FM by setting the receiver frequency near, but not on, the FM unmodulated-carrier frequency. An AM receiver has a narrowband filter with a passband of a few kilohertz. This gives a selectivity curve such as that shown in Fig. 10-5B. If the FM unmodulated-carrier frequency is near the *skirt* (sloping part) of the filter response, frequency modulation makes the signal move in and out of the passband. This causes the receiver output amplitude to vary, producing a fair reproduction of the original modulating waveform.
- *Phase-locked loop (PLL)*: If an FM signal is injected into a PLL, the loop produces an error voltage that is a duplicate of the modulating waveform. A *limiter* can be placed ahead of the PLL, so the signal passes through the limiter before it gets to the PLL. When this is done, the receiver does not respond to variations in the signal

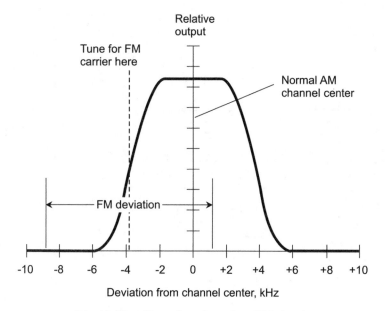

Fig. 10-5B.   Slope detection of an FM signal.

amplitude. Weak signals "break up" rather than fade in an FM receiver that employs limiting. A marginal FM signal sounds something like
the signal from a cell phone when the user is at the outer limit of the
cell range.

- *Discriminator*: This type of FM detector produces an output voltage
  that depends on the instantaneous signal frequency. When the signal
  is at the center of the passband, the output voltage is zero. If the instantaneous signal frequency falls below center, the output voltage
  becomes positive. If the frequency rises above center, the output
  becomes negative. A discriminator is sensitive to amplitude variations,
  but this can be overcome by a limiter stage placed ahead of it.

- *Ratio detector*: This FM detector is a discriminator with a built-in
  limiter. The original design was developed by RCA (Radio
  Corporation of America), and is still used today in some receivers.
  A simple ratio detector circuit is shown in Fig. 10-5C. The "balance"
  control is set for the best received signal quality.

DETECTION OF SSB

A single-sideband (SSB) signal is an AM signal with the carrier and one of
the sidebands removed. An incoming SSB signal can be combined with the

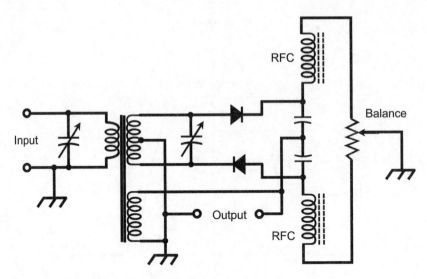

Fig. 10-5C. An FM ratio detector.

signal from an unmodulated LO, reproducing the original modulating data from the source transmitter. This is called *product detection*, and it is done at a single frequency, rather than at a variable frequency as in direct-conversion reception.

Two product-detector circuits are shown in Figs. 10-5D and E. At D, diodes are used, and there is thus no amplification. At E, a bipolar transistor is employed; this circuit provides some gain. The essential characteristic of either circuit is the generation of sum and difference frequency signals. (In this respect, product detectors are like mixers.) This is how the sideband energy is converted back into the modulating intelligence from the source.

PROBLEM 10-4
Suppose the transistor in the circuit of Fig. 10-5E suddenly stops amplifying, but still allows both the signal input and the LO input to pass through, and still mixes them to produce product detection. What will happen to the output signal?

SOLUTION 10-4
The output signal will still contain the same data as it had before the amplification failure, but it will be much weaker.

PROBLEM 10-5
Suppose, in the circuit of Fig. 10-5E, a PNP bipolar transistor is used instead of the NPN device shown. Suppose the characteristic curves of the PNP

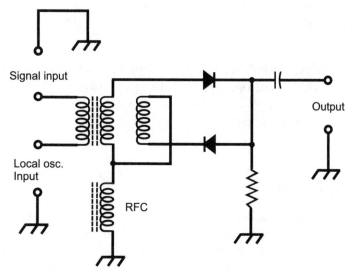

Fig. 10-5D. A product detector using two diodes.

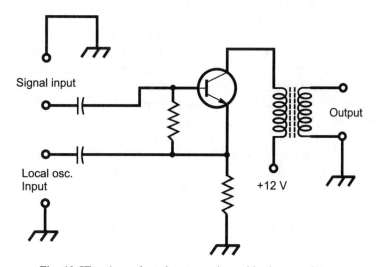

Fig. 10-5E. A product detector using a bipolar transistor.

device are the same as those for the NPN device. In what way must the circuit be changed so it will work with the PNP transistor?

SOLUTION 10-5
The polarity of the power-supply voltage at the terminal marked "+12 V" must be reversed. To reflect this, the terminal should be re-labeled so it says "−12 V." This negative voltage must be supplied with respect to ground.

Audio Stages

In a radio receiver, enhanced selectivity can be obtained by tailoring the audio frequency response in stages following the detector. The audio stages can also provide extra amplification. If a loudspeaker is used, the audio stages must provide enough output power to drive it. This is usually about 1 W.

AUDIO FILTERING

A voice signal requires a band of frequencies ranging from about 300 Hz to 3000 Hz in order to be conveyed in an intelligible manner. An *audio bandpass filter* with a passband of 300 Hz to 3000 Hz (which represents a bandwidth of 2700 Hz) can improve the quality of reception with some voice receivers. An ideal voice audio filter has little or no attenuation within the passband range, and high attenuation outside the range, and the skirts (dropoffs in the selectivity curve) are steep. This is called a *rectangular response*.

A CW Morse code signal requires only a few hundred hertz of bandwidth to be clearly read. Audio CW filters can narrow the response bandwidth to as little as 100 Hz. Passbands narrower than about 100 Hz produce *ringing*, degrading the quality of reception. The center frequency of a CW audio filter should be set to about 700 Hz, which represents a comfortable listening pitch for most people. Most CW audio filters have adjustable center frequencies.

An *audio notch filter* is a band-rejection filter with a sharp, narrow response. An interfering carrier, called a *heterodyne*, that produces a tone of constant frequency in the receiver output, can be cancelled out. Audio notch filters are tunable from roughly 300 Hz to 3000 Hz. Some sophisticated devices tune automatically; when a heterodyne appears and remains for a few hundredths of a second, the notch centers itself on the frequency of the heterodyne.

SQUELCHING

A *squelch* silences a receiver when no signal is present, allowing reception of signals when they appear. Most FM communications receivers use squelching systems. The squelch is normally closed when no signal is present, so no noise comes from the receiver. A signal opens the squelch if its amplitude exceeds the *squelch threshold*, which can be adjusted by the operator. When this happens, all signals and noise can be heard at the receiver output.

In some systems, the squelch will not open unless the signal has certain characteristics. The most common method of *selective squelching* uses AF tone generators. This can prevent unauthorized transmissions from accessing repeaters or being picked up by receivers. The squelch opens only for signals accompanied by a tone having the correct frequency, or by a sequence of tones having the correct frequencies in the correct order.

PROBLEM 10-6

Suppose a receiver's audio output circuits can pass a band of frequencies much greater than necessary in order to receive a voice. For example, suppose the circuits can pass frequencies from 20 Hz to 20,000 Hz. If such a receiver is used to receive a voice communications signal, what will happen?

SOLUTION 10-6

The receiver will work all right for this purpose under good conditions (no interference and low noise). But there will be severe interference if signals appear on nearby radio frequencies. In addition, if there is much external noise such as thunderstorm "static," more of this noise will get through along with the essential voice information, compared with a circuit that passes energy only in the range 300 to 3000 Hz.

Television Reception

A television (TV) receiver has a tunable front end, an oscillator and mixer, a set of IF amplifiers, a video demodulator, an audio demodulator and amplifier chain, a display with associated peripheral circuitry, and a loudspeaker.

FAST-SCAN TV

A receiver for analog *fast-scan television* (FSTV) is shown in simplified block form in Fig. 10-6. In standard American TV, there are 525 lines per frame, and 30 complete frames per second. As of this writing, the 525-line, 30-frame-per-second NTSC (National Television Standards Committee) scheme remains popular in the United States, but this is being replaced by digital television. Television broadcasts are made on various channels. Each NTSC channel is 6 MHz wide, including video and audio information.

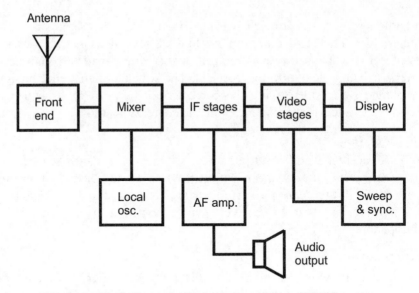

Fig. 10-6.   Block diagram of a fast-scan television (FSTV) receiver.

DIGITAL TELEVISION

Digital television is the transmission and reception of moving video images in digitized form. Communications satellites have proven popular for this purpose. Until the early 1990s, a satellite television installation required a dish antenna 2 or 3 meters in diameter. Digitization changed this situation. Engineers have managed to get the diameter of the receiving dish down to less than 1 meter.

A pioneer in digital TV was RCA (Radio Corporation of America), who developed the *Digital Satellite System* (DSS). An analog video signal is changed into digital pulses at the transmitting station by means of an A/D converter (ADC). The digital signal is amplified and sent up to a satellite. This signal is called the *uplink*. The satellite has a *transponder* that receives the signal, converts it to a different frequency, and retransmits it back toward the earth. The return signal is called the *downlink*. The downlink is picked up by a portable dish that can be placed on a balcony or patio, on a rooftop, or indoors near a window. A tuner circuit selects the channel that the subscriber wants to view. The digital signal is amplified and processed. It can then be changed back into analog form suitable for viewing on a conventional TV set by means of a *digital-to-analog converter* (D/A converter or DAC), as shown in Fig. 10-7. Alternatively, a special high-definition receiving set can be used. Programs are available that allow digital television signals to be viewed on

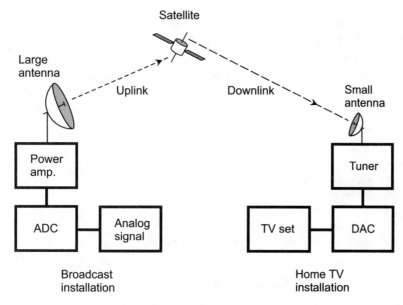

Fig. 10-7.   A complete digital TV link.

personal computers. The image detail, or *resolution*, is superior to that of NTSC television.

SLOW-SCAN TV

A *slow-scan television* (SSTV) communications station needs a transceiver with SSB capability, a standard NTSC TV set or personal computer with the appropriate software, a video camera, and a *scan converter*. The standard transmission rate is 8 frames per second. This reduces the required bandwidth to the essential audio voice frequency range of 300 Hz to 3000 Hz. That way, the SSTV signal can be sent directly over conventional SSB equipment without modification.

The scan converter consists of two data converters (one for receiving and the other for transmitting), a memory circuit, a tone generator, and a detector. Scan converters are commercially available. Computers can be programmed to perform this function. Amateur radio operators often build their own scan converters.

PROBLEM 10-7
Why are the uplink and downlink signals in Fig. 10-7 on different frequencies? Why can't the satellite receive a signal, amplify it on the same frequency, and retransmit it back to earth?

SOLUTION 10-7
The downlink signal, as it is transmitted from the satellite antenna, must not interfere with the received uplink signal arriving from the earth. If the two signals were on the same frequency, this interference would be practically impossible to prevent. Even when the downlink frequency is significantly different from the uplink frequency, isolation circuits are required to prevent the transmitted signal from desensitizing the receiver.

Specialized Modes

Some rather esoteric wireless communications techniques are effective under certain circumstances. Here are some examples.

DUAL-DIVERSITY RECEPTION

Dual-diversity reception, also known simply as *diversity reception*, is a scheme for reducing the effects of *fading* (fluctuations in received signal strength) in radio reception that occurs when signals from distant transmitters are returned to the earth from the ionosphere. This phenomenon is common at frequencies from 3 to 30 MHz, a range that is sometimes called the *shortwave band*. Two receivers are used; both are tuned to the same signal. They employ separate antennas, spaced several wavelengths apart. The outputs of the receivers are combined, in phase, at the input of a single audio amplifier (Fig. 10-8).

In dual-diversity reception, the receiver tuning is critical. Both receivers must be tuned to precisely the same frequency. In some installations, three or more antennas and receivers are employed. This provides even better immunity to fading, but it compounds the tuning difficulty and increases the expense.

SYNCHRONIZED COMMUNICATIONS

Digital signals require less bandwidth than analog signals to convey a given amount of information per unit time. *Synchronized communications* is a specialized digital mode in which the transmitter and receiver operate from a common time standard to optimize the amount of data that can be sent in a channel or band.

In synchronized digital communications, also called *coherent communications*, the receiver and transmitter operate in lock-step. The receiver evaluates

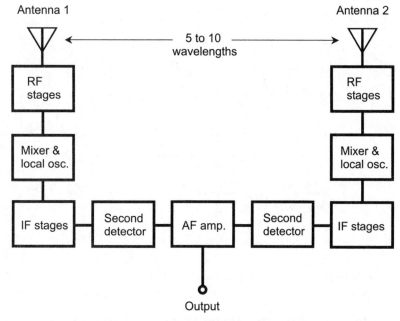

Fig. 10-8.   Dual-diversity reception uses two antennas and two identical radio receivers tuned to the same frequency.

each transmitted bit as a unit, for a block of time lasting from the bit's exact start to its exact finish. This makes it possible to use a receiving filter having extremely narrow bandwidth. The synchronization requires the use of an external time standard such as the broadcasts of the shortwave radio station WWV. *Frequency dividers* are employed to obtain the necessary synchronizing frequencies from the standard. A tone or pulse is generated in the receiver output for a particular bit if, but only if, the average signal voltage exceeds a certain value over the duration of that bit. False signals, such as might be caused by filter ringing, sferics, or other noise, are ignored because they do not produce sufficient average bit voltage.

Experiments with synchronized communications have shown that the improvement in S/N ratio, compared with nonsynchronized systems, is several decibels at low to moderate data speeds.

DIGITAL SIGNAL PROCESSING (DSP)

Digital signal processing (DSP) can extend the workable range of a communications circuit, because it allows reception under worse conditions than would be possible without it. Digital signal processing also improves the

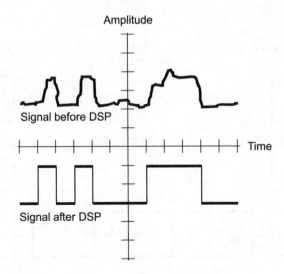

Fig. 10-9. Digital signal processing (DSP) can "clean up" a signal.

quality of fair signals, so that the receiving equipment or operator makes fewer errors. In circuits that use only digital modes, DSP can be used to "clean up" the signal. This also makes it possible to copy digital data over and over many times (that is, to produce multi-generation duplicates).

In DSP with analog modes such as SSB or SSTV, the signals are first changed into digital form using A/D conversion. Then the digital data is "cleaned up" so the pulse timing and amplitude adhere strictly to the set of technical standards (known as the *protocol*) for the type of digital data being used. Finally, the digital signal is changed back to the original voice or video using D/A conversion.

DSP minimizes noise and interference in a received digital signal as shown in Fig. 10-9. A hypothetical signal before DSP is shown at the top; the signal after processing is shown at the bottom. If the incoming signal is above a certain level for an interval of time, the DSP output is high (logic 1). If the level is below the critical point for a time interval, then the output is low (logic 0).

MULTIPLEXING

Signals in a communications channel or band can be intertwined in various ways. The most common methods are *frequency-division multiplexing* (FDM) and *time-division multiplexing* (TDM). Multiplexing requires an *encoder* at the transmitter and a *decoder* at the receiver.

In FDM, the channel is broken down into subchannels. The carrier frequencies of the signals are spaced so they do not overlap. Each signal is independent of the others. In TDM, signals are broken into segments by time, and then the segments are transferred in a rotating sequence. The receiver must be synchronized with the transmitter. They can both be clocked from a time standard such as WWV.

SPREAD SPECTRUM

In *spread-spectrum communications*, the main carrier frequency is rapidly varied independently of signal modulation, and the receiver is programmed to follow. As a result, the probability of *catastrophic interference*, in which one strong interfering signal can obliterate the desired signal, is near zero. It is difficult for unauthorized people to eavesdrop on a communication in progress.

Frequency-spreading functions can be complex and can be kept secret. If the operators of the transmitting station and all receiving stations do not divulge the function to anyone else, and if they do not tell anyone else about the existence of their communication, then no one else on the band will know the contact is taking place. It's possible for any number of receivers to intercept a spread-spectrum signal as long as the receiving operators know the frequency-spreading function. The use of such a function constitutes a form of *data encryption*.

During a spread-spectrum contact between a given transmitter and receiver, the operating frequency might fluctuate over a range of kilohertz, megahertz, or tens of megahertz. As a band becomes occupied with an increasing number of spread-spectrum signals, the overall noise level in the band appears to increase, as "heard" by a receiver tuned to a fixed frequency. There is a practical limit to the number of spread-spectrum contacts that a band can handle. This limit is about the same as it would be if all the signals were constant in frequency.

A common method of generating spread spectrum is *frequency hopping*. The transmitter has a list of channels that it follows in a certain order. The receiver must be programmed with this same list, in the same order, and must be synchronized with the transmitter. The *dwell time* is the length of time the transmitter remains on any given frequency between hops. The dwell time should be short enough so that a signal will not be noticed, and not cause interference, on any single frequency. There are numerous *dwell frequencies* so the signal energy is diluted to the extent that, if someone tunes to any frequency in the sequence, the signal is not noticeable.

Another method of obtaining spread spectrum, called *frequency sweeping*, is to frequency-modulate the main transmitted carrier with a waveform that guides it up and down over the assigned band. This FM is independent of signal intelligence. A receiver can intercept the signal if, but only if, its tuning varies according to the same waveform function, over the same band, at the same frequency, and in the same phase as the transmitter.

PROBLEM 10-8
What will happen if the timing between the transmitter and receiver is disrupted in a synchronized communications system?

SOLUTION 10-8
The enhanced S/N ratio will be lost. In some systems, if synchronization is disrupted, the system will revert to a standard, nonsynchronized mode (called *asynchronous communications*), so reception is still possible, although at a reduced level of performance.

Quiz

Refer to the text in this chapter if necessary. Answers are in the back of the book.

1. The extent to which a receiver can minimize variations in output but remain sensitive for signals ranging from weak to strong is called its
   (a) envelope detection rating.
   (b) DSP sensitivity.
   (c) dynamic range.
   (d) selectivity.

2. An asset of spread-spectrum communications is the fact that
   (a) it allows only for SSTV reception.
   (b) it cannot be received by more than one station at a time.
   (c) it is always a digital mode.
   (d) catastrophic interference is unlikely.

3. An AM receiver can receive an FM signal using
   (a) dual-diversity detection.
   (b) slope detection.
   (c) product detection.
   (d) envelope detection.

4. Digital signal processing can be used
   (a) as a detector in a crystal set.
   (b) to "clean up" a digital signal.
   (c) to reduce the S/N ratio in a superheterodyne receiver.
   (d) as the local oscillator in a mixer.

5. What is the function of a DAC?
   (a) It changes an analog signal into a digital one.
   (b) It serves as a detector for FM.
   (c) It operates as a slope detector for SSB.
   (d) None of the above.

6. A slow-scan television video receiver should have a passband approximately
   (a) 2.7 kHz wide.
   (b) 2.7 MHz wide.
   (c) 6 kHz wide.
   (d) 6 MHz wide.

7. The shortwave band covers a frequency range of approximately
   (a) 3 GHz to 30 GHz.
   (b) 300 MHz to 3 GHz.
   (c) 30 MHz to 300 MHz.
   (d) 3 MHz to 30 MHz.

8. A common time standard is necessary for the proper operation of
   (a) a synchronized communications system.
   (b) a superheterodyne receiver.
   (c) an FM receiver.
   (d) a ratio detector.

9. The diode in a crystal set should have
   (a) high resistance.
   (b) high inductance.
   (c) low capacitance.
   (d) low conductance.

10. A band-rejection filter with a sharp, narrow response, suitable for suppressing heterodynes in the audio stages of a receiver, is called
    (a) an envelope detector.
    (b) a discriminator.
    (c) a ratio detector.
    (d) an audio notch filter.

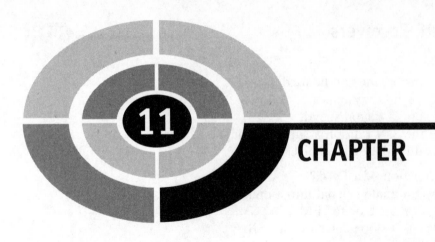

Telecommunications

The term *telecommunications* refers to the transfer of data between individuals, businesses, and/or governments. This can be done by *landline* (hard-wired systems), by *wireless* (EM waves), or by a combination of both.

Networks

The *Internet*, also known as *the Net*, is a worldwide system of interconnected computers. It was formed in the 1960s as *ARPAnet*, named after the Advanced Research Project Agency (ARPA) of the United States government.

PACKETS

When a data file or program is sent over the Net, it is divided into units called *packets* at the *source*, or transmitting computer. A packet consists of a *header*

Destination data

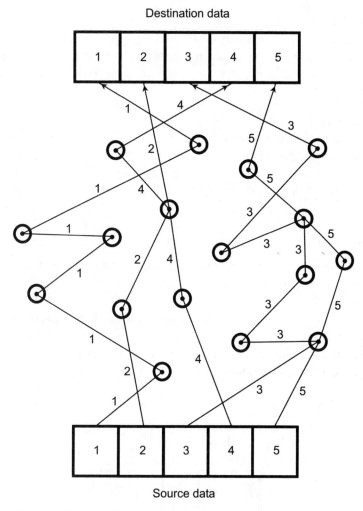

Fig. 11-1. Internet data flows in packets from the source to the destination.

followed by a certain number of data *bits* (binary digits) or *bytes* (groups of bits, usually 8 bits). Each packet is routed individually. The packets are reassembled at the *destination*, or receiving computer, into the original message.

Figure 11-1 is a simplified drawing of Internet data transfer for a hypothetical file containing 5 packets. Intermediate computers in the circuits, called *nodes*, are the black dots surrounded by circles. The file or program cannot be completely reconstructed until all the packets have arrived, and the destination computer has ensured that there are no errors.

PROBLEM 11-1

In the communication scenario of Fig. 11-1, suppose the number of intervening nodes (shown as circles with dots inside) between the source and destination were much greater for each transferred packet. What effect would this have?

SOLUTION 11-1

It would take somewhat longer for all the packets to reach the destination, and for the complete message to be assembled there. In the case of e-mail, for example, this would translate to a longer period of time between the transmission of the message and its appearance at the destination.

THE MODEM

The term *modem* is a contraction of *modulator/demodulator*. A modem interfaces a computer to a telephone line, cable network, optical fiber network, or radio transceiver.

Figure 11-2 is a block diagram of a modem suitable for interfacing a home or business computer with the telephone line. The modulator, or D/A converter, changes outgoing digital data into audio tones. The demodulator, or A/D converter, changes incoming audio tones into digital data. The audio

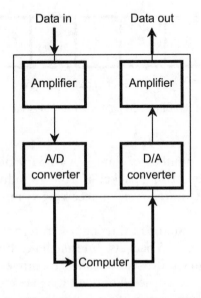

Fig. 11-2.   Block diagram of a telephone-line modem.

tones fall within the band of approximately 300 Hz to 3000 Hz, the same as that used for voice communications.

E-MAIL AND NEWSGROUPS

For some computer users, communication via Internet *electronic mail (e-mail)* and/or *newsgroups* has practically replaced the Postal Service. To use e-mail or newsgroups, everyone must have an *e-mail address*. An example is

sciencewriter@tahoe.com

The part of the address before the @ symbol is the *username*. The characters after the @ sign and before the dot represent the *domain name*. The three-letter abbreviation after the dot represents the *domain type*, for example, "com" for "commercial" or "org" for "organization."

INTERNET CONVERSATIONS

Computer users can carry on real-time (that is, "live") script conversations with other computer users over the Internet. This is called *Internet relay chat (IRC)*. It is also possible to digitize voice signals and transfer them by means of the Net. When the Net is not carrying very much data (called *light Net traffic*), such connections can be almost as good as those provided by telephone companies. But when there is *heavy Net traffic*, the quality is marginal to poor. When the volume of Net traffic is extreme, so-called *Internet telephone* connections may be impossible to establish or maintain.

GETTING INFORMATION

The Internet gets people in touch with billions of sources of information. Data is transferred among computers by means of a *file transfer protocol (FTP)*. When you use FTP, the files or programs at the remote computer become available to you, exactly as if they were stored in your own computer.

The *World Wide Web* (also called *WWW* or *the Web*) employs *hypertext*, a scheme of cross-referencing. Certain words, phrases, symbols (called *icons*), or images are highlighted or underlined. When you activate one of these *clickable links*, your computer goes to another document dealing with the same or a related subject.

LOCAL AND WIDE AREA NETWORKS

A *local area network (LAN)* is a group of interconnected computers located in each other's immediate vicinity. In a *client-server LAN* (Fig. 11-3A), there is one central computer called a *file server* to which smaller personal computers (labeled PC) are linked. In a *peer-to-peer LAN* (Fig. 11-3B), all of the computers are PCs with similar computing power, speed, and storage capacity. A peer-to-peer LAN offers greater privacy and user independence than a client-server LAN, but the peer-to-peer scheme is slower when all users share the same data.

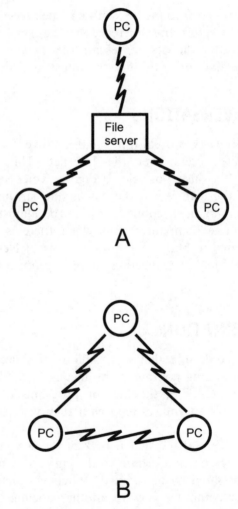

Fig. 11-3.   At A, a client-server LAN. At B, a peer-to-peer LAN.

A *wide area network (WAN)* is a group of computers linked over a large geographic region. Numerous LANs can be interconnected to form a WAN. The Internet is a good example of a WAN. Some corporations, universities, and government agencies operate their own WANs.

Satellites

A *communications satellite* system is like a gigantic cellular network in which the repeaters are in orbit around the earth, rather than at fixed locations on the surface. The end users can be in fixed, mobile, or portable locations.

GEOSTATIONARY SATELLITES

At an altitude of approximately 36,000 km, a satellite in a circular orbit takes one solar day to complete each revolution. If a satellite is placed in such an orbit over the equator, and if it revolves in the same direction as the earth rotates, it is a *geostationary satellite*. From the viewpoint of someone on the earth, a geostationary satellite stays in the same spot in the sky all the time. A single geostationary satellite can provide communications coverage over about 40% of the earth's surface. Three such satellites spaced 120° apart in longitude can provide coverage over the entire civilized world.

Earth-based stations can communicate using a single satellite only when the stations are both on a line of sight with the satellite. That means the "bird" must be above the horizon as "seen" from both surface locations. If two stations are nearly on opposite sides of the planet, they must operate through two geostationary satellites (Fig. 11-4).

LOW-EARTH-ORBIT (LEO) SATELLITES

A satellite in a geostationary orbit requires constant positional adjustment. Geostationary satellites are expensive to launch and maintain. There is a signal delay because of the path length. It takes high transmitter power and a precisely aimed dish antenna to communicate reliably. These problems with geostationary satellites have given rise to the *low-earth-orbit* (LEO) satellite concept.

In a LEO system, there are dozens of satellites spaced strategically around the globe in orbits a few hundred kilometers above the surface. The orbits

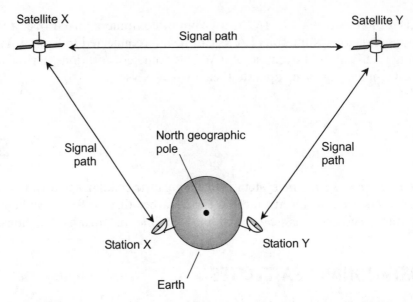

Fig. 11-4. A communications link using two geostationary satellites.

take the satellites near or over the geographic poles. Such an orbit, which is inclined at nearly 90° to the equator, is known as a *polar orbit*. The satellites relay messages among each other, and to and from end users. If there are enough satellites, any two end users can maintain constant contact.

A LEO satellite link is easier to use than a geostationary-satellite link. A simple, non-directional antenna will suffice, and it does not have to be aimed. The transmitter can reach the network using only a few watts of power. Signal-propagation delay, called *latency*, is much shorter than is the case with a link through a geostationary satellite system.

GLOBAL POSITIONING SYSTEM (GPS)

The *Global Positioning System* (GPS) is a network of satellites that allows users to determine their exact latitude, longitude and (if applicable) altitude above the surface or above sea level. The satellites transmit signals with extremely short wavelengths, comparable to those of a police radar. The signals are modulated with timing and identification codes.

A GPS receiver contains a computer that calculates the line-of-sight distances to four different satellites simultaneously, by comparing timing codes from signals arriving from each satellite. This computer can thus provide location data that is accurate to within a few meters of the user's actual position, based on the four distances.

PROBLEM 11-2
The average altitude of any satellite tends to change over time. Usually, the orbit decays (the satellite loses altitude). What will occur if this happens to a geostationary satellite and the altitude is not readjusted?

SOLUTION 11-2
The orbital period of any satellite gets shorter as the altitude decreases. If a geostationary satellite "falls" to an orbit slightly lower than the prescribed altitude, it speeds up, and therefore it begins to drift slowly from west to east as viewed from any fixed point on the earth's surface. The orbital period of any satellite gets longer as the altitude increases. If a geostationary satellite "rises" to an orbit slightly higher than the prescribed altitude, it begins to drift slowly from east to west as viewed from any fixed point on the earth's surface.

Personal Communications Systems

Personal communications systems (PCS) include cell phones, pagers, beepers, and all sorts of related paraphernalia. It seems that every day, some company comes out with a new device that can be used for private communications.

CELLULAR TELECOMMUNICATIONS

A *cellular telecommunications* system is a network of *repeaters*, also known as *base stations*, allowing portable or mobile radio transceivers to be used as telephone sets. A *cell* is the coverage zone of a base station.

If a cell phone set is in a fixed location such as a residence, then communication to and from that phone set usually takes place through a single cell. If the cell phone set is in a moving vehicle such as a car or boat, it goes from cell to cell (Fig. 11-5A). In this drawing, the cellular repeaters are represented by small dots, the coverage cells are represented by circles, and the path of the vehicle is represented by the heavy curve. All the base stations are connected to the telephone system by wires, microwave links, or fiberoptic cables.

A notebook computer can, in theory, be connected to a cell phone set with a specialized modem (Fig. 11-5B) for access to the Internet. As of this writing, most cell phone sets are not very well equipped to work this way,

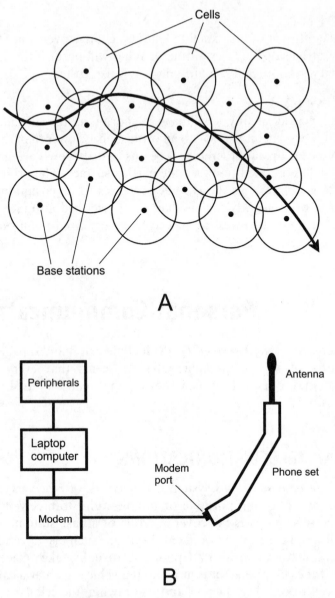

Fig. 11-5.   At A, the cellular telecommunications concept. At B, the connection of a notebook
computer to a cell phone set.

and when access can be obtained, the connection is usually slow. But wireless
networks are appearing across the United States and other developed
countries, and these allow high-speed portable Internet access with the use of
modest antennas, independent of the telephone system.

PROBLEM 11-3

Refer to Fig. 11-5A. Suppose you are driving along a roadway, your engine dies, and you are forced to pull off the road and stop at a point located at the fringes of the coverage zones of two base stations. (Such a point would be indicated in Fig. 11-5A by two circles just touching each other, but not overlapping.) You make a call for a tow truck. The connection is made, but the signal has an annoying flutter. What is the reason for this flutter?

SOLUTION 11-3

The cellular system is attempting to complete the connection using both of the base stations for which you are at the fringes. The repeaters are competing. Your signal is going through one and then the other, and then the one again, over and over. This is common and nearly always produces an echoing or fluttering effect. Almost everyone who has used a cell phone has experienced this!

WIRELESS LOCAL LOOP

A *wireless local loop* (WLL) is similar to a cellular system. Telephone sets and data terminals are linked into the system by means of radio transceivers. An example is shown in Fig. 11-6. Heavy lines represent wire connections, and thin lines represent wireless links. The small circles and ellipses represent subscribers. Each subscriber can use a telephone set (small circle) or a personal computer (small ellipse) equipped with a modem. Several telephone sets and/or computers can be connected together by a system of wires confined to the house or business.

PAGERS

A simple *pager* or *beeper* employs a small, battery-powered radio receiver. Transmitters are located in various places throughout a city, county, or telephone area code district. The receiver picks up a signal that causes the unit to display a series of numerals representing the sender's telephone number.

A pager system equipped with *voice mail* allows the sender to leave a brief spoken message following the beep. A pager receiver equipped to receive e-mail resembles a handheld computer or small calculator. When the unit emits a sequence of beeps, the user looks at the screen to read messages. Messages can be stored for later retrieval or transfer to a laptop or desktop computer.

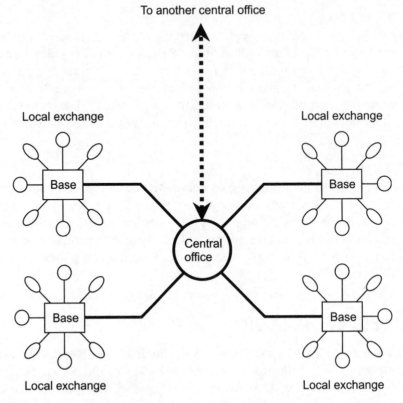

Fig. 11-6. Wireless local loop (WLL) telephone systems. Heavy lines represent wire links, and thin lines represent wireless links. Small circles represent telephone sets, and small ellipses represent computers.

Some pagers can send e-mail messages as well as receive them. This is done using a system similar to a cellular telephone network. The pager contains a small radio transmitter with an attached whip antenna. Two notes of caution:

- The use of wireless e-mail transmitters is generally forbidden on aircraft.
- Reading or writing e-mail should never be done while driving a motor vehicle.

WIRELESS FAX

Facsimile (fax) is a method of sending and receiving non-moving images, as you've already learned. Fax can be done over wireless networks, as well as

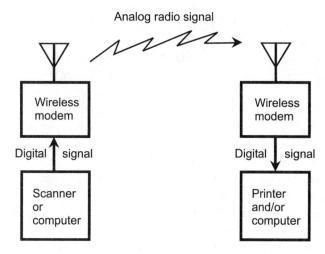

Analog radio signal

Fig. 11-7.   Transmission of a wireless fax signal.

over the telephone. Figure 11-7 shows the transmission of a wireless fax signal. Only the sending part of the fax machine is shown at the source (left), and only the receiving part is shown at the destination (right). A complete fax installation has a transmitter and a receiver, allowing for two-way exchange of images.

To send a fax, a page of printed material is placed in an *optical scanner*. This device converts the image into binary digital signals (1 and 0). The output of the scanner is sent to a modem that converts the binary digital data into an analog signal suitable for transmission. At the destination, the analog signals are converted back into digital pulses like those produced by the optical scanner at the source. These pulses are routed to a printer, data terminal, or computer.

Hobby Communications

Hobby communications includes *shortwave listening* (SWLing), the *Citizens Radio Service*, and *amateur radio*.

SHORTWAVE LISTENING

A *high-frequency* (HF) radio communications receiver is sometimes called a *shortwave receiver*. Most of these radios function at all frequencies from 3 MHz through 30 MHz. Some also work in the standard AM broadcast band at 535 kHz to 1.605 MHz, and in the spectrum between the AM broadcast band and 3 MHz.

Technically, the shortwave or HF band extends from 3 MHz to 30 MHz. The range of frequencies from 300 kHz to 3 MHz is called the *medium-wave* or *medium-frequency (MF)* band, and the range from 30 kHz to 300 kHz is called the *longwave* or *low-frequency (LF)* band. A receiver that can "hear" from 30 kHz to 30 MHz is called a *general-coverage receiver*.

Some receivers can "hear" in the ultra-high-frequency (UHF) range from 300 MHz to 3 GHz, as well as at lower frequencies all the way down to a few kilohertz. These deluxe radios are called *all-wave receivers*.

In the United States, a shortwave listener (or "SWLer") need not obtain a license to receive signals. But in some countries, a license may be needed to own a radio receiver. Shortwave listeners often get interested enough in communications to obtain amateur radio licenses, allowing them to transmit signals on various bands of frequencies throughout the radio spectrum.

CITIZENS RADIO SERVICE

The Citizens Radio Service, also known as *Citizens Band* (CB), is a public radio communications and control service. The most familiar form is *Class D*, which operates on 40 discrete channels near 27 MHz (11 meters) in the HF band.

A 40-channel, 12-watt transceiver is the basic radio for class-D fixed-station operation. It employs the SSB voice mode and operates from the standard 117-volt utility circuit. Mobile transceivers also run 12 watts SSB, and operate from 13.7-volt vehicle batteries. The power connection in a mobile installation should be made directly to the battery. "Cigarette-lighter adapters" are not recommended for this purpose.

The organization of *Radio Emergency Associated Communications Teams*, usually called by its acronym, *REACT*, is a worldwide group of radio communications operators. They provide assistance to authorities in disaster areas. On the Class-D band, the emergency channel at 27.065 MHz (channel 9) is monitored by REACT operators.

The *General Mobile Radio Service*, or GMRS, operates at frequencies between 460 and 470 MHz using *Class A* Citizens Band. The maximum communications range between two individual transceivers in Class A is about 40 miles. Communications beyond 40 miles is done using repeaters.

Some classes of CB operation require government licenses, while others do not. For the latest regulations, check at an electronics store that sells CB equipment.

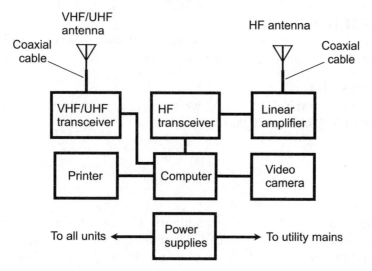

Fig. 11-8.   A basic amateur radio station.

AMATEUR RADIO

A fixed amateur (or "ham") radio station has several components (Fig. 11-8). A computer can be used to communicate by means of *packet radio* with other hams who own computers. The station can be equipped for online telephone (landline) services. The computer can control the antennas for the station, and can keep a log of all stations that have been contacted. Most modern transceivers can be operated by computer, either locally or by remote control over the radio or landline.

Mobile ham radio equipment is operated in a moving vehicle such as a car, truck, train, boat, or airplane. Mobile equipment is generally more compact than fixed-station apparatus. In addition, mobile gear is designed to withstand large changes in temperature and humidity, as well as mechanical vibration.

Portable ham radio equipment is almost always operated from battery power, and can be set up and dismantled quickly. Some portable equipment can be operated while being physically carried around; an example is the *handy-talkie* (HT) or *walkie-talkie*. Portable equipment must withstand vibration, temperature and humidity extremes, and prolonged use.

All amateur radio operation requires licensing. The *American Radio Relay League (ARRL)* is the most recognized organization of ham radio operators in the world. They are eager to help anyone who wants to get a ham radio license. They can be reached on the Web at *www.arrl.org*.

PROBLEM 11-4
One of the amateur radio bands covers a frequency range of 3500 kHz to 4000 kHz. In what part of the radio spectrum is this?

SOLUTION 11-4
This band is in the HF range, which covers frequencies from 3 MHz to 30 MHz (3000 kHz to 30,000 kHz).

Lightning

Lightning is a hazard to radio amateurs, CB operators, and shortwave listeners. An outdoor antenna can accumulate a large electrostatic charge during a thundershower. In case of a nearby strike, the *electromagnetic pulse (EMP)* can produce a massive surge of current in an antenna. A direct lightning hit on an antenna produces a catastrophic current and voltage surge that can start fires and electrocute people. Dangerous voltage "spikes" can occur on utility mains and telephone lines.

THE NATURE OF LIGHTNING

A lightning "bolt," technically called a *stroke*, lasts for a small fraction of a second. But tremendous power is liberated in that short time because of the high current and voltage. There are four types of lightning:

- Lightning that occurs within a single cloud (*intracloud*), shown at A in Fig. 11-9.
- Lightning in which the major current pulse flows from a cloud to the earth's surface (*cloud-to-ground*), shown at B in Fig. 11-9.
- Lightning that occurs between two clouds (*intercloud*), shown at C in Fig. 11-9.
- Lightning in which the major current pulse flows from the earth's surface to a cloud (*ground-to-cloud*), shown at D in Fig. 11-9.

Cloud-to-ground and ground-to-cloud lightning present the greatest danger to electronics hobbyists and equipment. Intracloud or intercloud lightning can cause an EMP sufficient to damage sensitive apparatus, particularly if the antenna has long elements connected directly to a radio. Some radio hams and SWLers use *longwire antennas* that can accumulate a large voltage and pick up a substantial EMP even from a lightning stroke several kilometers away.

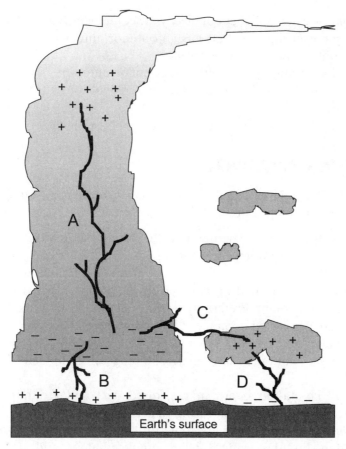

Fig. 11-9.   Four types of lightning stroke: intracloud (A), cloud-to-ground (B), intercloud (C), and ground-to-cloud (D).

PROTECTING YOURSELF

The following precautions are recommended near and in thundershowers. These measures will not guarantee immunity, but they minimize the danger:

- Stay indoors, or inside a metal enclosure such as a car, bus, or train.
- Stay away from windows.
- If it is not possible to get indoors, find a low-lying spot on the ground, such as a ditch or ravine, and squat down with your feet close together until the threat has passed.
- Avoid lone trees or other isolated, tall objects such as utility poles or flagpoles.

- Avoid electric appliances or electronic equipment that makes use of the utility power lines, or that has an outdoor antenna.
- Stay out of the shower or bathtub.
- Avoid swimming pools, either indoors or outdoors.
- Do not use hard-wired telephones or personal computers with modems connected.

PROTECTING HARDWARE

Precautions that minimize the risk of damage to electronic equipment (but not guarantee immunity) are as follows:

- Never operate, or experiment with, a ham, shortwave, or CB radio station when lightning is occurring near your location.
- When the station is not in use, disconnect all antennas and ground all feed line conductors to a good electrical ground other than the utility power-line ground. Preferably the lines should be left outside the building and connected to an earth ground several feet away from the building.
- When the station is not in use, unplug all equipment from utility outlets.
- When the station is not in use, disconnect and ground all antenna rotator cables and other wiring that leads outdoors.
- *Lightning arrestors* provide some protection from electrostatic-charge buildup, but they cannot offer complete safety, and should not be relied upon for routine protection.
- *Lightning rods* reduce (but don't eliminate) the chance of a direct hit, but they should not be used as an excuse to neglect the other precautions.
- Power line *transient suppressors* (also called "surge protectors") reduce computer "glitches" and can sometimes protect sensitive components in a power supply, but they should not be used as an excuse to neglect the other precautions.
- The antenna mast or tower should be connected to an earth ground using heavy-gauge wire or braid. Several parallel lengths of American Wire Gauge (AWG) No. 8 ground wire, run in a straight line from the mast or tower to ground, form an adequate conductor.
- Other secondary protection devices are advertised in electronics-related and radio-related magazines.

For more information, consult a competent communications engineer. If you are in doubt about the fire safety of an electronic installation, consult your local fire inspector. It's better to hassle a bureaucrat than to have your house burn down.

PROBLEM 11-5
How can a lightning strike on a power line cause damage to home entertainment equipment that is plugged into wall outlets, even when the equipment is switched off?

SOLUTION 11-5
The induced surge, which can attain peak values of several thousand volts, arcs across the power-switch contacts. When this voltage appears across the power-supply transformer primary, it produces a current surge. This current in turn produces a high-voltage surge across the transformer secondary. This "spike" is transmitted to the internal components of the system.

Security and Privacy

People are concerned about the security and privacy of information exchanged over telecommunications systems, because this information can be misused by unauthorized persons.

WIRELESS VERSUS WIRED

Wireless eavesdropping differs from *wiretapping* in two fundamental ways. First, eavesdropping is easier to do in wireless systems than in hard-wired systems. Second, eavesdropping of a wireless link is impossible to physically detect, but a tap can be found in a hard-wired system.

If any portion of a communications link occurs over a wireless link, then an eavesdropping receiver can be positioned within range of the RF transmitting antenna (Fig. 11-10) and the signals intercepted. The existence of the *wireless tap* has no effect on the behavior of equipment in the system.

LEVELS OF SECURITY

There are four levels of telecommunications security, ranging from level 0 (no security) to the most secure connections technology allows.

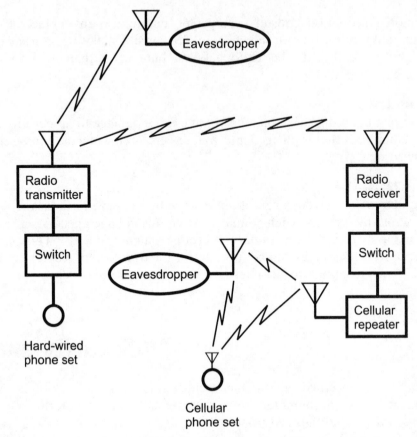

Fig. 11-10. Eavesdropping on RF links in a telephone system. Heavy, straight lines represent wires or cables; zig-zags represent RF signals.

No security (level 0): In a *level-0-secure communications system,* anyone can eavesdrop on a connection at any time, provided they are willing to spend the money and/or time to obtain the necessary equipment. Examples of level-0 links are amateur radio and Citizens' Band (CB) voice communications. The lack of privacy is compounded by the fact that if someone is eavesdropping, none of the communicating parties can detect the intrusion.

Wire-equivalent security (level 1): An end-to-end hard-wired connection requires considerable effort to tap, and sensitive detection apparatus can usually reveal the existence of any wiretap. A special code called a *cipher* may be used. The intent of the cipher is to conceal the data content from everyone except authorized persons or entities. A *level-1-secure communications system* must have certain characteristics in order to be effective and practical:

- The cost must be affordable.
- The system must be reasonably safe for transactions such as credit-card purchases.
- When network usage is heavy, the degree of privacy afforded to each subscriber should not decrease, relative to the case when network usage is light.
- The cipher, if used, should be unbreakable for at least 12 months, and preferably for 24 months or more.
- Encryption technology, if used, should be updated at least every 12 months, and preferably every six months.

Security for commercial transactions (level 2): Some financial and business data warrants protection beyond wire-equivalent. Many companies and individuals refuse to transfer money by electronic means because they fear criminals will gain access to an account. In a *level-2-secure communications system*, the encryption used in commercial transactions should be such that it would take a hacker at least 10 years, and preferably 20 years or more, to break the cipher. The technology should be updated at least every 10 years, but preferably every 3 to 5 years.

Mil-spec security (level 3): Security to military specifications (mil spec) involves the most sophisticated encryption available. Technologically advanced countries, and entities with economic power, have an advantage. The encryption in a *level-3-secure system* should be such that engineers believe it would take a hacker at least 20 years, and preferably 40 years or more, to break the cipher. The technology should be updated as often as economics allow.

EXTENT OF ENCRYPTION

Security and privacy are obtained by *digital encryption*. The idea is to render signals readable only to receivers with the necessary *decryption key*.

For level-1 security, encryption is required only for the wireless portion(s) of the circuit. The cipher should be changed at regular intervals to keep it "fresh." The block diagram of Fig. 11-11A shows wireless-only encryption for a hypothetical cellular telephone connection.

For security at levels 2 and 3, *end-to-end encryption* is necessary. The signal is encrypted at all intermediate points, even those for which signals are transmitted by wire or cable. Figure 11-11B shows this scheme in place for the same hypothetical cellular connection as depicted at A.

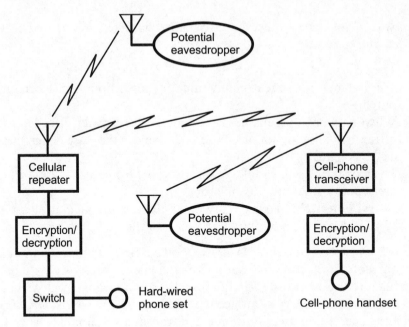

Fig. 11-11A. Wireless-only encryption. Heavy, straight lines represent wires or cables; zig-zags represent RF signals.

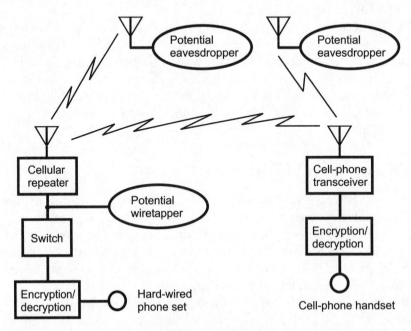

Fig. 11-11B. End-to-end encryption. Heavy, straight lines represent wires or cables; zig-zags represent RF signals.

SECURITY WITH CORDLESS PHONES

If someone knows the frequencies X and Y at which a base unit and cordless handset operate, and if that person is determined to eavesdrop on conversations that take place over that system, it is possible to place a *wireless tap* on the line. The conversation can be intercepted and recorded at a remote site (Fig. 11-12). It is more difficult to design and construct a wireless tap for a multiple-channel cordless set, as compared with a single-channel set. But if all the channel frequencies are known, they can be monitored with a scanning receiver.

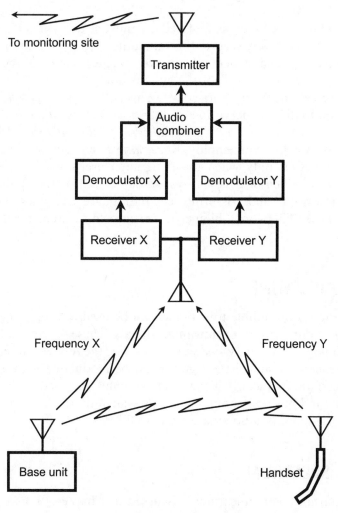

Fig. 11-12.   Wireless tapping of a cordless telephone.

PART 3 **Wireless Electronics**

If there is any security concern about using a cordless phone, a hard-wired phone set should be used.

SECURITY WITH CELL PHONES

Cellular telephones are, in effect, long-range cordless phones. The wider extent of geographical coverage increases the risk of eavesdropping and unauthorized use. Digital encryption of the actual conversation – both ends of it – is an effective way to maintain privacy and security of cellular communications. Anything less than this leaves a conversation subject to interception.

Access and privacy codes, as well as data, must be encrypted if a cell-phone system is to be maximally secure. If an unauthorized person knows the code with which a cell phone set accesses the system (the "name" of the set), other sets can be programmed to fool the system into "thinking" the bogus sets belong to the user of the authorized set. This is known as *cell-phone cloning*.

In addition to digital encryption of data, *user identification* (user ID) must be employed. The simplest is a *personal identification number* (PIN). More sophisticated systems can employ *voice-pattern recognition*, in which the phone set functions only when the designated user's voice speaks into it. *Hand-print recognition* or *iris-print recognition* can also be employed. Voice-pattern, hand-print, and iris-print recognition are examples of *biometric security systems*. The word "biometric" means "measurement of biological characteristics."

TONE SQUELCHING

Subaudible tones or audible tone bursts can be used to keep a receiver from responding to unnecessary or unwanted signals. These schemes all fall into a set of technologies known as *tone squelching*. For the receiver squelch to open in the presence of a signal, the signal must be modulated with a sine wave having a certain frequency. Tone-burst systems are used by some police departments and business communications systems in the U.S. Tone squelching is popular among radio amateurs as well.

SPREAD SPECTRUM

Spread-spectrum communications, discussed in Chapter 10, can offer some security. Special receivers are necessary to receive signals in this mode.

A conventional receiver or scanner cannot intercept spread-spectrum signals unless the receiver follows exactly along with the changes in the signal frequency. The frequency-variation sequence can be complicated. In its most sophisticated form, is almost as good as digital encryption for security purposes. Spread-spectrum can be used in addition to digital encryption and other security measures to provide a "layered" security scheme.

AUDIO SCRAMBLING

Audio scrambling is an older technology that is less sophisticated than modern encryption and more subject to interception. An amateur-radio single-sideband (SSB) transmitter and receiver can demonstrate basic scrambling (Fig. 11-13). The transmitter is connected to a *dummy antenna* and is set for USB operation. The receiver is set for LSB reception, and is tuned about 3000 Hz higher than the transmitter. The receiver is connected indirectly to the transmitter through an *attenuator* that allows only a small signal from the transmitter to get to the receiver antenna terminals.

Suppose the suppressed-carrier frequency of the USB transmitter is 3.800000 MHz. This produces a spectral output similar to that shown in the

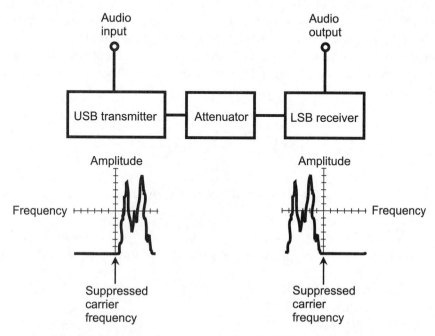

Fig. 11-13.   A simple voice scrambling and descrambling circuit.

graph on the left. Each horizontal division represents 500 Hz, and each vertical division represents 5 dB. The receiver is tuned to 3.803000 MHz. The signal energy from the transmitter therefore falls into the receiver passband, as shown by the graph on the right, but the audio frequencies are inverted within the 3000-Hz voice passband. To demonstrate unscrambling, the audio output is tape-recorded, and the recorded signal is applied to the input. The "upside-down" signal is rendered "right-side-up" again.

Sophisticated audio scramblers split up the audio passband into two or more sub-passbands, inverting some or all of the segments, and rearranging them frequency-wise within the main passband. But audio scrambling is an analog mode, and cannot provide the levels of protection available with digital encryption.

PROBLEM 11-6
What is the difference between encoding and encryption?

SOLUTION 11-6
Encoding is the translation of data into an orderly form (that is, a code) according to a widely known protocol. The oldest example is the Morse code, in which each character is assigned a unique set of short and long pulses called *dots* and *dashes*. No attempt is made to conceal the information from anyone; the encoding is a matter of expediency. Encryption, in contrast, is the alteration of data using a cipher, with the intent to conceal the content from everyone except authorized persons or entities. In order to retrieve encrypted data, the receiving machine or operator must have a decryption key.

Quiz

Refer to the text in this chapter if necessary. Answers are in the back of the book.

1. A local area network in which all of the terminals are computers having similar power, speed, and storage capacity is known as
   (a) a peer-to-peer LAN.
   (b) a client-server LAN.
   (c) a hard-wired LAN.
   (d) a wireless LAN.

2. In end-to-end encryption,
   (a) the signal is encrypted only where it is transmitted by wireless means.
   (b) the signal is encrypted only where it is transmitted by wire or cable.
   (c) the signal is encrypted at all points between the source and the destination.
   (d) Any of the above.

3. Fill in the blank in the following sentence to make it true: "Lightning can cause damage to electronic equipment _____ as a result of the EMP produced by the lightning stroke."
   (a) on board geostationary satellites
   (b) thousands of kilometers away
   (c) connected to an outdoor antenna
   (d) disconnected from all antennas and power sources

4. In order for a cell phone connection to take place, a mobile or portable phone set
   (a) must be outside the range of all repeaters.
   (b) must be inside the coverage range of a repeater.
   (c) must be inside the coverage range of two or more repeaters.
   (d) must be hard wired to the telephone line.

5. Which of the following types of security is biometric?
   (a) Spread spectrum
   (b) Digital signal processing
   (c) Audio scrambling
   (d) None of the above

6. When a signal is altered with the intent of concealing the information from everyone except certain authorized persons or entities, the process is called
   (a) encoding.
   (b) digital signal processing.
   (c) analog-to-digital conversion.
   (d) encryption.

7. Which of the following types of encryption is the most secure?
   (a) Morse code
   (b) Amplitude modulation
   (c) Digital encryption
   (d) Audio scrambling

8. Intermediate computers, through which data goes on its way from the source to the destination on the Internet, are called
   (a) packets.
   (b) nodes.
   (c) bits.
   (d) bytes.

9. Internet telephone connections may be difficult or impossible to establish
   (a) when the volume of Internet traffic is extremely light.
   (b) when the volume of Internet traffic is extremely heavy.
   (c) using broadband connections.
   (d) when the computers are too fast.

10. In an Internet e-mail address, the username
   (a) appears before the @ symbol.
   (b) appears after the @ symbol but before the dot.
   (c) appears after the dot.
   (d) can appear anywhere.

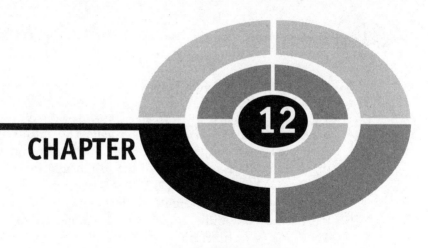

CHAPTER 12

Antennas

There are hundreds of different kinds of antennas in wireless electronic systems. In this chapter we'll learn how antennas work, and examine a few common types.

Radiation Resistance

All antennas have an interesting property called *radiation resistance*. A "thought experiment" can help you get an idea of it. Imagine that a plain resistor is substituted for a transmitting antenna that doesn't have any capacitance or inductance. Suppose the radio transmitter behaves in the same manner when connected to the resistor, as it does when connected to the antenna. For any antenna operating at a specific frequency, there exists a specific resistance, in ohms, for which this can be done. This is the radiation resistance (R_R) of the antenna at the frequency in question.

DETERMINING FACTORS

Suppose a thin, straight vertical wire is placed over a gigantic sheet of metal that forms a *perfectly conducting ground*. The output of a radio transmitter is connected between the sheet of metal and the bottom end of the wire. When this is done, R_R is a function of the wire height in wavelengths. The function is shown as a graph in Fig. 12-1A.

Suppose a thin, straight, lossless wire is placed in empty space, and the output of a radio transmitter is connected to the center of the wire. Then R_R is a function of the overall conductor length in wavelengths (Fig. 12-1B).

The wavelength of a radio signal depends on the frequency. If the frequency in megahertz (MHz) is represented by the lowercase italic letter f_{MHz}, and wavelength in meters (m) is represented by the lowercase italic Greek letter lambda with the subscript m (λ_m), then:

$$\lambda_m = 300/f_{MHz}$$
$$\text{and}$$
$$f_{MHz} = 300/\lambda_m$$

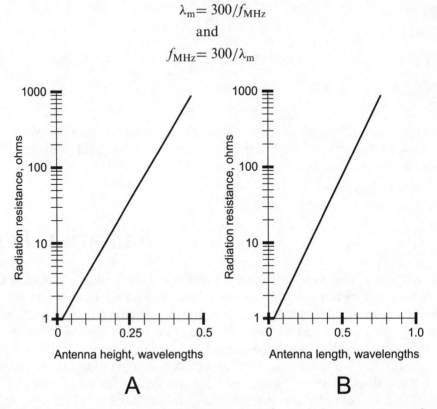

Fig. 12-1. Approximate values of radiation resistance for vertical antennas over perfectly conducting ground (A) and for center-fed antennas in free space (B).

If feet (ft) are used for the wavelength rather than meters, then the formulas are:

$$\lambda_{ft} = 984/f_{MHz}$$
$$\text{and}$$
$$f_{MHz} = 984/\lambda_{ft}$$

where λ_{ft} represents the wavelength in feet.

ANTENNA EFFICIENCY

Efficiency is important in a transmitting antenna system. It's best if a radio transmitting antenna has a large R_R, because radiation resistance appears in series with *loss resistance* (R_L). The real earth ground is never perfectly conducting, antenna wires always have some resistance, and there are always objects near antennas that can absorb EM energy and prevent it from going out to distant receivers.

The efficiency of an antenna (Eff) is given by:

$$\text{Eff} = R_R/(R_R+R_L)$$

This is the ratio of the radiation resistance to the total antenna system resistance. As a percentage, efficiency ($\text{Eff}_\%$) is calculated this way:

$$\text{Eff}_\% = 100R_R/(R_R+R_L)$$

PROBLEM 12-1
Suppose the radiation resistance of a vertical antenna is $37\,\Omega$, but the ground and surrounding objects produce a loss resistance of $111\,\Omega$. (In regions where the soil is dry or sandy, the loss resistance can be this high, or higher.) What is the antenna efficiency in percent? What does this figure mean in terms of transmitted RF signal power lost?

SOLUTION 12-1
In this case, $R_R = 37\,\Omega$ and $R_L = 111\,\Omega$. Plugging these numbers into the formula for antenna efficiency in percent, we get this:

$$
\begin{aligned}
\text{Eff}_\% &= 100 \times 37/(37 + 111) \\
&= 100 \times 37/148 \\
&= 100 \times 0.25 \\
&= 25\%
\end{aligned}
$$

This means that only 25% of the RF power or energy supplied to the antenna is radiated into space where it can eventually be picked up by distant receivers. The other 75% is dissipated as heat in the loss resistance.

Half-Wave Antennas

Antennas measuring a half wavelength ($\lambda/2$) are common, and perform well when they are properly designed and well placed. The most popular basic types are discussed in the following paragraphs.

FORMULAS

A half wavelength in free space is given by the equation:

$$L_m = 150/f_{MHz}$$

where L_m is the linear distance in meters, and f_{MHz} is the frequency in megahertz. A half wavelength in feet is given by:

$$L_{ft} = 492/f_{MHz}$$

For ordinary wire, the results as obtained above must be multiplied by a *velocity factor*, *v*, of 0.95 (95%). For metal tubing, *v* can range down to approximately 0.90 (90%). The reason for this is the fact that RF waves travel only 90% to 95% as fast along an antenna element than they do through empty space.

OPEN DIPOLE

An *open dipole* or *doublet* is a half-wavelength radiator fed at the center. Each "leg" of the antenna is therefore 1/4 wavelength ($\lambda/4$) long, as shown in Fig. 12-2A. For a straight wire radiator, the length L_m, in meters, at a design frequency f_{MHz}, in megahertz, for a practical half-wave dipole is approximately:

$$L_m = 143/f_{MHz}$$

The length in feet is approximately:

$$L_{ft} = 467/f_{MHz}$$

These values assume $v = 0.95$.

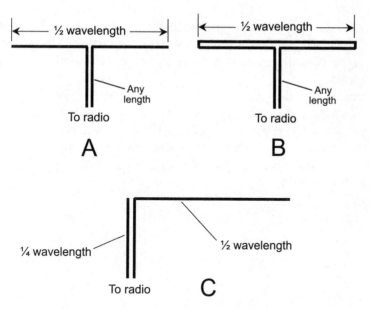

Fig. 12-2. Basic half-wave antennas. At A, the dipole antenna. At B, the folded-dipole antenna. At C, the zepp antenna.

PROBLEM 12-2

What is the end-to-end length of a half-wavelength open dipole antenna, constructed of wire, for use at a frequency of 7.100 MHz? Express the answer in meters, and also in feet.

SOLUTION 12-2

To calculate the end-to-end length of the antenna, "plug in the numbers" to the formulas:

$$L_m = 143/7.100$$
$$= 20.1\,\text{m}$$
$$\text{and}$$
$$L_{ft} = 467/7.100$$
$$= 65.8\,\text{ft}$$

FOLDED DIPOLE

A *folded dipole antenna* is a half-wavelength, center-fed antenna constructed of two parallel wires with their ends connected together (Fig. 12-2B). The folded dipole is used with parallel-wire transmission lines, in applications where gain and directivity are not especially important. Perhaps you have seen this type of antenna sold for use as an indoor receiving antenna for FM broadcast.

The formulas for the end-to-end length of a folded dipole are the same as those for an open dipole constructed of small-diameter metal tubing. The folded portions of the antenna don't operate as a transmission line, but instead, they radiate signals. Some trial-and-error is required when designing a folded-dipole antenna, but the following formulas are good approximations, based on $v = 0.925$:

$$L_m = 139/f_{MHz}$$
$$\text{and}$$
$$L_{ft} = 455/f_{MHz}$$

ZEPP

A *zeppelin antenna*, also called a *zepp*, is a half-wavelength radiator, fed at one end with a quarter-wavelength section of parallel-wire line (Fig. 12-2C). A zepp is efficient at all harmonics of the design frequency, as well as at the design frequency itself. The length of the radiating element is critical for this type of antenna. If it is not exactly a half wavelength (calculated using the formulas incorporating the velocity factor, as described above) or a whole-number multiple thereof, the feed line will not work properly. Feed lines are discussed later in this chapter.

Quarter-Wave Antennas

Antennas measuring a quarter wavelength ($\lambda/4$) are also common in wireless communications. They work well if they are properly designed and employ good ground systems.

FORMULAS

A *quarter wavelength* in free space is related to frequency by the equation:

$$H_m = 75.0/f_{MHz}$$

where H_m represents a quarter wavelength in meters, and f_{MHz} represents the frequency in megahertz. If the height of the antenna is expressed in feet, then the formula is:

$$H_{ft} = 246/f_{MHz}$$

If v is the velocity factor in a medium other than free space, then:

$$H_m = 75.0v/f_{MHz}$$
$$\text{and}$$
$$H_{ft} = 246v/f_{MHz}$$

For a typical wire conductor, $v = 0.95$ (95%); for metal tubing, v can range down to approximately 0.90 (90%). For a vertical antenna constructed from aluminum tubing such that $v = 0.92$, for example, the formulas are

$$H_m = 69.0/f_{MHz}$$
$$\text{and}$$
$$H_{ft} = 226/f_{MHz}$$

A quarter-wave antenna must be operated against a good RF ground. That means the ground losses must be kept to a minimum. Otherwise, the efficiency suffers.

GROUND-MOUNTED VERTICAL

The simplest possible vertical antenna is a quarter-wave radiator mounted at ground level. The radiator is fed with coaxial cable. The center conductor of the cable is connected to the base of the radiator, and the shield of the cable is connected to an RF ground system that includes the earth itself.

At some frequencies, the height of a quarter-wave vertical is unmanageable unless *inductive loading* is used to reduce the physical length of the radiator. A quarter-wave vertical antenna can be made resonant on several frequencies by the use of multiple *loading coils*, or by inserting parallel-resonant LC circuits known as *traps* at specific points along the radiator. The design of this sort of antenna is tricky, because adjusting any of its resonant frequencies changes all the others.

Unless an extensive *ground radial* system is installed, a ground-mounted vertical is inefficient. Inductive loading worsens this situation, because it involves the use of a physically shortened radiator whose R_R is lower than that of a full-height quarter-wave radiator. Another problem is the fact that vertical antennas often receive more human-made noise than horizontal antennas.

In addition, the EM fields from ground-mounted transmitting antennas are more likely to cause interference to nearby consumer electronic devices, than are the EM fields from antennas installed at a height.

GROUND-PLANE ANTENNA

A *ground-plane antenna* is a vertical radiator operated against a system of quarter-wave radials and elevated at least 1/4 wavelength above the earth's surface. The radiator itself should be tuned to resonance at the desired operating frequency.

When a ground plane is elevated at least 1/4 wavelength above the surface, only three or four radials are necessary to obtain a low-loss RF ground system. The radials extend outward from the base of the antenna at an angle between 0° and 45° with respect to the horizon. Figure 12-3A illustrates a ground-plane antenna in which the radiating element measures λ/4. A ground-plane antenna should be fed with coaxial cable.

The radials in a ground-plane antenna can extend straight down as a quarter-wave, cylindrical *sleeve* surrounding, and concentric with, the

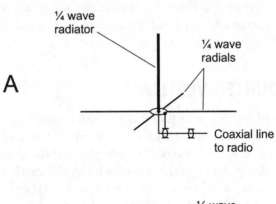

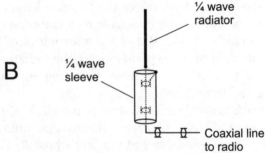

Fig. 12-3.   At A, a basic ground-plane antenna with a quarter-wave radiating element and quarter-wave radials. At B, a coaxial antenna.

coaxial-cable feed line. This is a modified ground-plane antenna known as a *coaxial antenna* (Fig. 12-3B).

PROBLEM 12-3
What is the height of a quarter-wavelength vertical antenna, constructed of aluminum tubing, for use at a frequency of 50.1 MHz? Express the answer in meters, and also in feet. Assume the velocity factor is equal to 0.900.

SOLUTION 12-3
"Plug in" the numbers to the formulas given above. This yields the following results:

$$H_m = 75v/f_{MHz}$$
$$= 75.0 \times 0.900/50.1$$
$$= 1.35\,m$$
$$\text{and}$$
$$H_{ft} = 246v/f_{MHz}$$
$$= 246 \times 0.900/50.1$$
$$= 4.42\,ft$$

Loop Antennas

Any receiving or transmitting antenna, consisting of one or more turns of wire forming a DC short circuit, is a *loop antenna*. Loops can be categorized as either small or large.

SMALL LOOP

A *small loop antenna* has a circumference of less than 0.1λ, and is suitable for receiving, but not for transmitting unless extraordinary measures are taken to minimize the resistance in the loop conductor and associated components. A small loop is the least responsive along its axis (the line running perpendicular to the plane containing the loop itself), and is most responsive in the plane containing the loop. A capacitor can be connected in series or parallel with the loop to provide a resonant response. An example is shown in Fig. 12-4.

Small loops are useful for *radio direction finding* (RDF), and also for reducing interference caused by human-made noise or strong local signals. The null along the axis is sharp and deep. As such a loop is rotated, there is a

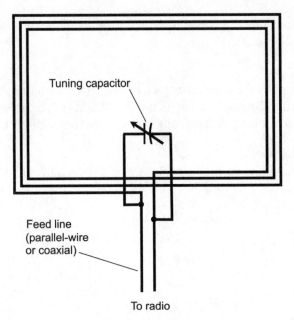

Fig. 12-4. A small loop antenna with a capacitor for adjusting the resonant frequency.

sharp decrease in the received signal strength when the axis of the loop points directly toward a nearby transmitting antenna or source of interference.

LOOPSTICK

For receiving applications at frequencies up to approximately 20 MHz, a *loopstick antenna* is sometimes used. This antenna, a variant of the small loop, consists of a coil of insulated or enameled wire, wound on a rod-shaped, powdered-iron core. A series or parallel capacitor, in conjunction with the coil, forms a tuned circuit.

A loopstick displays directional characteristics similar to those of the small loop antenna shown in Fig. 12-4. The sensitivity is maximum off the sides of the coil, and a sharp null occurs off the ends (that is, along the coil axis). This null can be used to minimize interference from local signals and from human-made sources of noise.

LARGE LOOP

If a loop has a circumference greater than 0.1 wavelength, it is classified as a *large loop antenna*. A large loop usually has a circumference of either 0.5λ or 1.0λ, and is resonant all by itself, without a tuning capacitor.

A large loop can be used for transmitting as well as for receiving. The maximum radiation and response for a half-wavelength loop occurs in the plane containing the loop. For this reason, a half-wavelength loop can be oriented with its axis oriented vertically, so it radiates and receives fairly well in all horizontal directions. The maximum radiation/response for a full-wavelength loop occurs along the axis. This type of loop is usually erected with its axis oriented horizontally.

For a large, square, full-wavelength loop made of wire, with each side a quarter-wavelength long, the circumference is given approximately by these formulas:

$$C_m = 306/f_{MHz}$$

and

$$C_{ft} = 1005/f_{MHz}$$

These are the values for meters and feet, respectively.

If you're astute, you'll notice something strange about these formulas. They suggest that, when an antenna is bent into a full-wavelength loop, its physical length is longer than an equivalent, straight-line wavelength in free space, as if the velocity factor is greater than 1! The reason for this is beyond the scope of this book, and involves the way the EM field around a full-wavelength loop interacts with itself. Rest assured that it doesn't mean the EM fields travel around such a loop faster than the speed of light, which is the same as the speed of all EM fields in free space.

PROBLEM 12-4

What is the circumference of a full-wavelength loop made of wire, designed for a resonant frequency of 14.0 MHz? Express the answer in meters, and also in feet.

SOLUTION 12-4

Using the above formulas, "plug in" the numbers to get these results:

$$C_m = 306/f_{MHz}$$
$$= 306/14.0$$
$$= 21.9 \, m$$

and

$$C_{ft} = 1005/f_{MHz}$$
$$= 1005/14.0$$
$$= 71.8 \, ft$$

Ground Systems

Unbalanced antenna systems, such as a quarter-wave vertical, require low-loss RF ground systems. Most balanced systems, such as a half-wave dipole, do not. A low-loss direct-current (DC) ground, also called an *electrical ground*, is advisable for any antenna system.

ELECTRICAL GROUNDING

Electrical grounding is important for personal safety. It can help protect equipment from damage if lightning strikes nearby. It minimizes the chances for *electromagnetic interference* (EMI) to and from the equipment.

There's an unwritten commandment in electrical and electronics engineering: "Never touch two grounds at the same time." This refers to the tendency for a potential difference to exist between two supposedly neutral points, in apparent defiance of logic. Experienced people have received severe electrical shocks because they forgot this rule. It's a good idea to wear electrically insulating gloves and rubber-soled shoes when working with ground systems of any kind.

Some appliances have 3-wire cords. One wire is connected to the "common" or "ground" part of the hardware, and leads to a D-shaped or U-shaped prong in the plug. This ground prong should never be defeated, because such modification can result in dangerous voltages appearing on exposed metal surfaces.

PROBLEM 12-5

What might cause a potential difference to exist between the grounded wire of a 3-wire electrical cord and a copper cold-water pipe in your house?

SOLUTION 12-5

Unless the copper pipe is connected by a wire directly to the grounded wires in the electrical system, a considerable AC voltage can exist between them. This voltage is induced by *electromagnetic coupling* between the wiring and the plumbing. Electrical currents can also exist in the earth itself, giving rise to potential differences between points in the soil located more than a few meters apart. Even if a cold-water pipe runs directly into the earth and all splices are metal-to-metal, an AC voltage can exist between the pipe and the ground wire of the household electrical system. This voltage can be high enough to cause a lethal electrical shock.

RF GROUNDING

A good *RF ground* system can help minimize EMI. Figure 12-5 shows a proper RF ground scheme (A) and an improper one (B). In a good RF ground system, each device is connected to a common *ground bus*, which in turn runs to the earth ground by means of a single conductor. This conductor should be as short as possible. A poor ground system contains *ground loops* that increase the susceptibility of equipment to EMI. Long ground wires and ground loops can act as antennas at RF, no matter how good the connection to the earth.

Ground radials measure 1/4 wavelength or more. They run outward from the base of a surface-mounted vertical antenna, and are connected to the shield of the coaxial feed line. The radials can be installed on the surface, or buried under the earth. The greater the number of radials of a given length in a surface-mounted vertical antenna system, the better the antenna will work. Also, the longer the radials for a given number, the better.

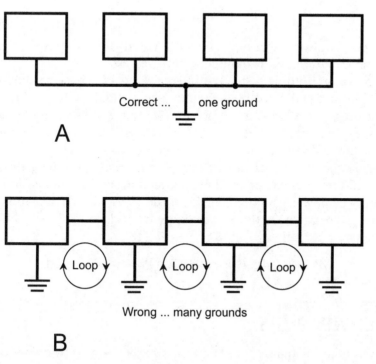

Fig. 12-5.   At A, the correct method for grounding multiple units. At B, an incorrect method creates RF ground loops.

As we have seen, if the base of a vertical antenna is 1/4 wavelength above the surface or higher, only three or four 1/4-wave radials are necessary to obtain a good RF ground.

A *counterpoise* is another means of obtaining an RF ground or ground plane without a direct earth-ground connection. A grid of wires, a screen, or a metal sheet is placed above the surface and oriented horizontally, to provide capacitive coupling to the earth. This minimizes RF ground loss. Ideally, the radius of a counterpoise should be at least 1/4 wavelength at the lowest operating frequency.

Gain and Directivity

The *power gain* of a transmitting antenna is the ratio of the maximum *effective radiated power* (ERP) to the actual RF power applied at the feed point. Power gain is expressed in *decibels* (dB). If the ERP is P_{ERP} watts and the applied power is P watts, then:

$$\text{Power Gain (dB)} = 10 \log_{10}(P_{\text{ERP}}/P)$$

Power gain is always measured in the favored direction(s) of an antenna, where the radiation and response are the greatest.

For power gain to be defined, a *reference antenna* must be chosen with a power gain that is defined by default as being equal to 0 dB. This reference antenna is usually a half-wave dipole in free space. Power-gain figures taken with respect to a dipole (in its favored directions) are expressed in units called dBd (decibels with respect to dipole). Alternatively, the reference antenna for power-gain measurements can be an *isotropic radiator*, which theoretically radiates and receives equally well in all directions in three dimensions. In this case, units of power gain are called dBi (decibels with respect to isotropic). For any given antenna, the power gains in dBd and dBi are different by approximately 2.15 dB:

$$\text{Power Gain (dBi)} = 2.15 + \text{Power Gain (dBd)}$$

DIRECTIVITY PLOTS

Antenna radiation and response patterns are represented by polar-coordinate plots such as those shown in Fig. 12-6. The location of the antenna is assumed to be at the center, or *origin*, of the graph. The greater the radiation

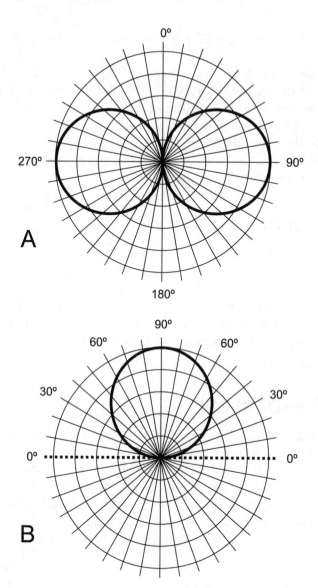

Fig. 12-6.   Directivity plots for a dipole antenna. At A, in the horizontal plane or H-plane; at B, in the elevation plane or E-plane.

or reception capability of the antenna in a certain direction, the farther from the origin the points on the chart are plotted.

A dipole antenna, oriented horizontally so that its conductor runs in a north-south direction, has a *horizontal-plane* (H-plane) directional pattern similar to that in Fig. 12-6A. In this illustration, the axis of the wire runs along the line connecting 0° and 180°. The angles correspond to directions of the compass, where $0° =$ north, $90° =$ east, $180° =$ south, and $270° =$ west.

The *elevation-plane* (E-plane) pattern of a dipole antenna depends on the height of the antenna above ground. With the dipole oriented so that its conductor runs perpendicular to the page and the antenna 1/4 wavelength above ground, the E-plane antenna pattern resembles the graph shown at Fig. 12-6B. In this case the angles represent degrees above the horizon, and the dashed line represents the horizon. (The lower half of the coordinate plane represents points below ground.)

FORWARD GAIN

The *forward gain* of an antenna is the ratio, in decibels, of the ERP in the main lobe relative to the ERP from a dipole antenna in its favored directions.

Some antennas manufactured from multiple elements of metal tubing can have forward gain figures exceeding 20 dBd. At microwave frequencies, large dish antennas can have forward gain upwards of 35 dBd. In general, as the wavelength decreases (the frequency gets higher), it becomes easier to obtain high forward gain figures.

FRONT-TO-BACK RATIO

The *front-to-back ratio* of a *unidirectional* (one-way) antenna, abbreviated f/b, is a decibel (dB) comparison of the ERP in the center of the main lobe relative to the ERP in the direction 180° opposite the center of the main lobe. Figure 12-7 shows a hypothetical H-plane directivity plot for a unidirectional antenna pointed north. The f/b ratio is found by comparing the ERP between north (0°) and south (180°). If each pair of adjacent, concentric circles in this graph represents a difference of 5 dB, then the f/b ratio in this case is 15 dB.

FRONT-TO-SIDE RATIO

The *front-to-side ratio* of an antenna, abbreviated f/s, is another expression of the directivity of an antenna system. The f/s ratio is also expressed in decibels (dB). The ERP in the center of the main lobe is compared with the ERP in the centers of the *side lobes* (the ones at right angles to the favored direction). An example is shown in Fig. 12-7. The f/s ratios are found, in this case, by comparing the ERP levels between north and east (right-hand f/s), or between north and west (left-hand f/s). The right-hand and left-hand f/s ratios

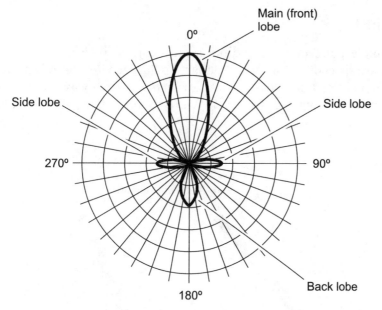

Fig. 12-7.   Directivity plot for a hypothetical antenna. Front-to-back and front-to-side ratios can be determined from such a graph.

are theoretically the same in most antennas. In practice, however, they are often different because electrically conducting objects near the antenna distort the pattern.

PROBLEM 12-6
Suppose that each pair of adjacent, concentric circles in Fig. 12-7 represents a difference of 10 dB. What is the f/s ratio for the antenna?

SOLUTION 12-6
The centers of the side lobes are about 3.5 circles inward from the center of the main lobe. Therefore, the f/s ratio in this example is approximately 35 dB.

Phased Arrays

A *phased array* uses two or more *driven elements* (elements directly connected to the feed line) to produce gain in some directions at the expense of other directions.

CONCEPT

Two simple pairs of phased dipoles are depicted in Fig. 12-8. At A, the dipoles are spaced λ/4 wavelength apart and are fed 90° out of phase, resulting in a unidirectional pattern. A *bidirectional* (two-way) pattern can be obtained by spacing the dipoles 1.0λ apart and feeding them in phase, as shown at B.

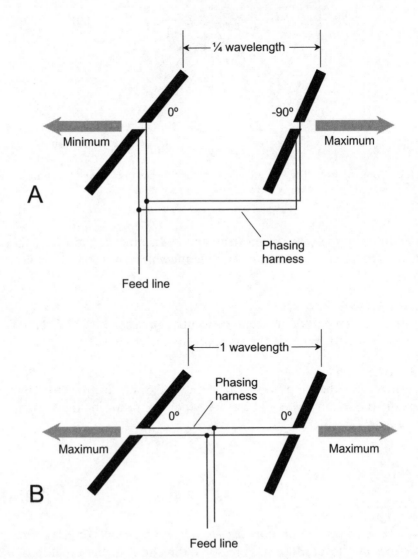

Fig. 12-8. At A, a unidirectional phased system. At B, a bidirectional phased system.

For proper operation, the lengths of the phasing sections are critical. In the system shown at Fig. 12-8A, the electrical distance between the junction of the two transmission-line sections and the right-hand antenna should be $\lambda/4$ longer than the distance between the junction and the left-hand antenna. In the system shown at Fig. 12-8B, the electrical distance between the junction of the two transmission-line sections and the right-hand antenna should be the same as the distance between the junction and the left-hand antenna. An electrical quarter wavelength is determined according to the following formulas:

$$Q_m = 75.0v/f_{MHz}$$

and

$$Q_{ft} = 246v/f_{MHz}$$

where Q_m is the length in meters, Q_{ft} is the length in feet, v is the velocity factor of the transmission line used for the phasing lines (called the *phasing harness*), and f_{MHz} is the frequency in megahertz.

Phased arrays can have fixed or steerable directional patterns. The pair of phased dipoles shown in Fig. 12-8A can, if the wavelength is short enough to allow construction from metal tubing, be mounted on a rotator for 360° directional adjustability. With phased vertical antennas, the relative signal phase can be varied, and the directional pattern thereby adjusted.

LONGWIRE ANTENNA

There is disagreement among antenna enthusiasts concerning how long a wire must be to qualify for *longwire* status. Some people say 1.0λ is long enough; others insist on 2.0λ or more. Some folks casually call random-length, end-fed wires shorter than 1.0λ "longwires," but such antennas are more appropriately called *random wires*.

In order to be a true longwire, an antenna must be straight, and it must be long enough to produce substantial gain over a dipole. The gain is a function of the length of the antenna: the longer the wire, the greater the gain. A straight longwire measuring several wavelengths can produce several dBd of gain. The radiation pattern is complicated, with multiple *major lobes* and *minor lobes*. The major lobes, where maximum radiation and response occur, are nearly in line with the wire. Minor lobes appear in many different directions. The pattern can change dramatically with only a small change in wire length.

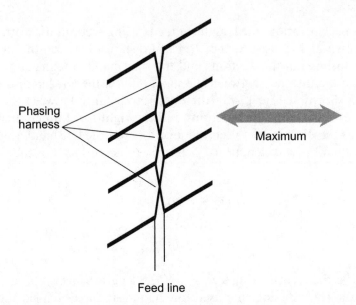

Fig. 12-9. A broadside array. The elements are all fed in phase. Maximum radiation and response occur perpendicular to the plane containing the elements.

BROADSIDE ARRAY

Figure 12-9 shows the geometric arrangement of a *broadside array*. The driven elements can each consist of a single radiator, as shown in the figure, or they can consist of complex systems with individual directive properties. If a reflecting screen is placed behind the array of dipoles in Fig. 12-9, the system becomes a *billboard antenna*.

The directional properties of a broadside array depend on the number of elements, whether or not the elements have gain themselves, and on the spacing among the elements. In general, the larger the number of elements, the greater the gain.

END-FIRE ARRAY

A typical *end-fire array* consists of two parallel half-wave dipoles fed $90°$ out of phase and spaced $\lambda/4$ apart. This produces a unidirectional radiation pattern. Alternatively, the two elements can be driven in phase and spaced at a separation of 1.0λ. This results in a bidirectional radiation pattern. In the phasing harness, the two branches of transmission line must be cut to precisely the correct lengths, and the velocity factor of the line must be known and be taken into account. The phased dipole systems of Fig. 12-8 are both end-fire arrays.

PROBLEM 12-7

What is the length of a quarter-wave transmission line, used as a phasing harness in an antenna similar to that shown in Fig. 12-8A, if the velocity factor of the line is 0.800 and the frequency is 18.1 MHz?

SOLUTION 12-7

Use the formulas and "plug in" the numbers to get the following:

$$Q_m = 75.0v/f_{MHz}$$

$$= 75.0 \times 0.800/18.1$$

$$= 3.31 \, m$$

and

$$Q_{ft} = 246v/f_{MHz}$$

$$= 246 \times 0.800/18.1$$

$$= 10.9 \, ft$$

Parasitic Arrays

Parasitic arrays are used at frequencies from about 2 MHz up to several giga-hertz for obtaining directivity and forward gain. The two simplest and most common examples are the *Yagi antenna* and the *quad antenna*.

CONCEPT

A *parasitic element* is an electrical conductor that comprises an important part of an antenna system, but that is not directly connected to the feed line. Parasitic elements operate by means of EM interaction with the driven element. When gain is produced in the direction of the parasitic element, the element is called a *director*. When gain is produced in the direction opposite the parasitic element, the element is called a *reflector*. Directors are typically a few percent shorter than the driven element; reflectors are a few percent longer.

YAGI ANTENNA

The *Yagi antenna*, sometimes called a *beam antenna*, is an array of parallel, straight elements. ("Yagi" was the name of one of the Japanese inventors of this antenna.) A two-element Yagi is formed by placing a director or a reflector parallel to, and a specific distance away from, a single driven element.

The optimum spacing for a driven-element/director Yagi is 0.1λ to 0.2λ, with the director tuned to a frequency 5% to 10% higher than the resonant frequency of the driven element. The optimum spacing for a driven-element/ reflector Yagi is 0.15λ to 0.2λ, with the reflector tuned to a frequency 5% to 10% lower than the resonant frequency of the driven element. The gain of a well-designed *2-element Yagi* is approximately 5 dBd.

A *3-element Yagi* with one director and one reflector, along with the driven element, increases the gain and f/b ratio compared with a 2-element Yagi. An optimally designed 3-element Yagi has approximately 7 dBd gain. An example is shown in Fig. 12-10. (This is a conceptual drawing, but not a design blueprint.)

The gain and f/b ratio figures for a Yagi increase as elements are added. This is usually done by placing extra directors in front of a three-element Yagi. Each director is slightly shorter than its predecessor. The design and construction of Yagi antennas having numerous elements is a sophisticated business. As the number of elements increases, the optimum antenna dimensions become more and more intricate.

QUAD ANTENNA

A *quad antenna* operates according to the same principles as the Yagi, except full-wavelength loops are used instead of straight half-wavelength elements.

A *2-element quad* can consist of a driven element and a reflector, or it can have a driven element and a director. A *3-element quad* has one driven element, one director, and one reflector. The director is tuned to a frequency 3% to 5% higher than the resonant frequency of the driven element. The reflector is tuned to a frequency 3% to 5% lower than the resonant frequency of the driven element.

Additional directors can be added to the basic three-element quad design to form quads having any desired numbers of elements. The gain increases as the number of elements increases. Each succeeding director is slightly shorter than its predecessor. Alternatively, the additional directors can be half-wave, straight elements similar to those of a multi-element Yagi,

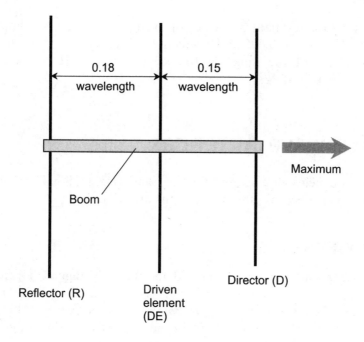

Length DE = 470/f
Length D = 425/f
Length R = 510/f

where f = operating frequency, MHz

Fig. 12-10.   A 3-element Yagi antenna. Dimensions are discussed in the text.

instead of loops. Such an antenna is called a *quagi*. Long quad and quagi antennas are physically practical only at frequencies above about 100 MHz.

PROBLEM 12-8

Suppose a 2-element Yagi antenna, having a driven element and one director, is designed. Its driven element is cut for a half-wave resonance at a frequency of 10.1 MHz. What is the approximate range of frequencies for which the director should be trimmed?

SOLUTION 12-8

The director in a 2-element Yagi should be trimmed for half-wave resonance at a frequency 5% to 10% higher than that of the driven element. This means the half-wave resonant frequency of the director should be between 1.05

and 1.10 times the resonant frequency of the director. The lower limit of this range is $10.1 \times 1.05 = 10.605$ MHz, which can be rounded off to 10.6 MHz. The upper limit of the range is $10.1 \times 1.10 = 11.11$ MHz, which can be rounded off to 11.1 MHz.

UHF and Microwave Antennas

At radio frequencies above approximately 300 MHz, high-gain antennas are reasonable in size because the wavelengths are short.

DISH ANTENNA

A *dish antenna* must be correctly shaped and precisely aligned. The most efficient shape is a *paraboloidal reflector*. This is a section of a *paraboloid*, which is the result of the rotation of a *parabola* around its axis in 3-dimensional space. A somewhat less precise, but in most cases workable, alternative is the *spherical reflector*, whose shape is a section of a sphere.

The feed system of a dish antenna usually consists of a coaxial cable or *waveguide* (a hollow conduit-like structure through which EM fields can propagate) from the receiver and/or transmitter, and a small antenna at the *focal point*. In some dish antenna systems, a small receiving preamplifier and transmitting unit are placed at the focal point, and are operated by remote control through coaxial cables. These systems constitute *conventional dish feed* (Fig. 12-11A). Alternatively, the antenna or preamplifier/transmitter can be placed behind the main dish, and the signals reflected from a convex "EM mirror" located at the focal point. This is *Cassegrain dish feed*, and is shown in Fig. 12-11B.

The larger the diameter of the reflector in wavelengths, the greater is the gain and the narrower is the main lobe. A dish antenna must be at least several wavelengths in diameter for proper operation. The reflecting element can be sheet metal, a screen, or a wire mesh. If a screen or mesh is used, the spacing between the wires must be a small fraction of a wavelength.

HELICAL ANTENNA

A *helical antenna* is a high-gain, unidirectional antenna with a shape something like a loosely wound air coil or spring, often in conjunction with a

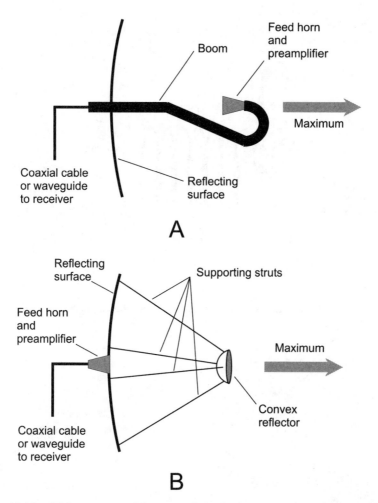

Fig. 12-11.   Dish antennas with conventional feed (A) and Cassegrain feed (B).

flat reflecting screen. Figure 12-12 is a functional diagram. A helical antenna can be used over a rather wide band of frequencies above a specific lower limit. The reflector diameter should be at least 0.8λ at the lowest operating frequency. The radius of the helix should be approximately 0.17λ at the center of the intended operating frequency range. The longitudinal spacing between turns should be approximately 0.25λ in the center of the operating frequency range. The overall length of the helix should be at least 1.0λ at the lowest operating frequency.

A helical antenna can provide about 15 dBd forward gain. Groups of helical antennas, all placed side-by-side in a square matrix with a single large reflector, are common in space communications systems.

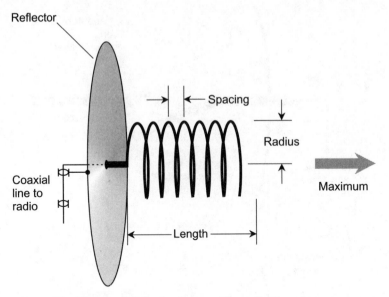

Fig. 12-12.   A helical antenna with a reflecting element.

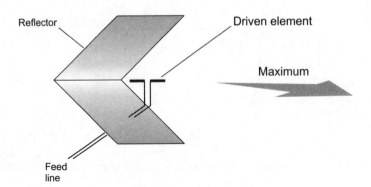

Fig. 12-13.   A corner-reflector antenna, used at ultra-high and microwave frequencies.

CORNER-REFLECTOR ANTENNA

A *corner reflector*, employed with a half-wave driven element, is illustrated in Fig. 12-13. This provides modest gain over a half-wave dipole. The reflector is wire mesh, screen, or sheet metal. Corner reflectors are used for television (TV) reception and satellite communications. Several half-wave dipoles can be fed in phase and placed along a common axis with a single, elongated reflector, forming a *collinear corner-reflector array*.

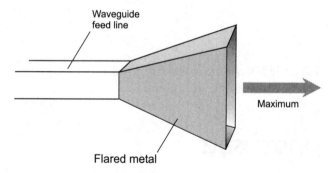

Waveguide
feed line

Maximum

Flared metal

Fig. 12-14.   A horn antenna, used at microwave frequencies.

HORN ANTENNA

There are several different configurations of the *horn antenna*; they all look similar. Figure 12-14 is a representative drawing. This antenna provides a unidirectional radiation and response pattern, with the favored direction coincident with the opening of the horn. The feed line is a waveguide that joins the antenna at the narrowest point (throat) of the horn.

Horns are not often used by themselves; they are commonly used to feed large dish antennas. The horn is aimed back at the center of the dish reflector. This minimizes extraneous radiation and response to and from the dish.

PROBLEM 12-9
What is the minimum overall length a helical antenna should have for a design frequency of 900 MHz?

SOLUTION 12-9
The minimum length should be 1.0λ in free space. Recalling the formulas above, and "plugging in" the numbers, gives us these results in meters and feet:

$$\lambda_m = 300/f_{MHz}$$
$$= 300/900$$
$$= 0.333 \, m$$

and

$$\lambda_{ft} = 984/f_{MHz}$$
$$= 984/900$$
$$= 1.09 \, ft$$

Feed Lines

A *feed line*, also called a *transmission line*, transfers energy between wireless communications hardware and an antenna.

BASIC CHARACTERISTICS

Feed lines can be categorized as either unbalanced or balanced. Examples of *unbalanced feed line* include coaxial cables and waveguides. Parallel-wire cables such as *twinlead* (also called *TV ribbon*) and *ladder line* are examples of *balanced feed line*.

All feed lines exhibit a *characteristic impedance* (Z_0) that depends on the physical construction of the line. This value, expressed in ohms (Ω), is the ratio of the RF voltage (in volts) to the RF current (in amperes) when there are no standing waves on the line, that is, when the current and voltage are in the same ratio at all points along the line.

Coaxial cable typically has Z_0 between $50\,\Omega$ and $100\,\Omega$. Twinlead is available with characteristic impedances ranging from approximately $75\,\Omega$ to $300\,\Omega$. Ladder line has Z_0 values between $300\,\Omega$ and $600\,\Omega$, depending on the spacing between the conductors, and also on the type of dielectric material that is used to keep the spacing constant between the conductors.

WAVEGUIDE

Waveguides are used at UHF and microwave frequencies, that is, at frequencies above about $300\,\text{MHz}$. A typical waveguide is a hollow metal pipe, usually having a rectangular or circular cross section. In order to efficiently propagate an EM field, a *rectangular waveguide* must have cross-sectional dimensions measuring at least 0.5λ, and preferably more than 0.7λ. A *circular waveguide* should be at least 0.6λ in diameter, and preferably 0.7λ or more.

The Z_0 of a waveguide varies with the frequency. In this sense, it differs from coaxial or parallel-wire lines, whose Z_0 values are independent of the frequency. Waveguides must be kept clean and dry inside, and free of obstructions. This can pose maintenance problems; water and condensation must be prevented from forming while spiders and insects must be kept out.

STANDING WAVES

Standing waves are voltage and current variations that exist on an RF transmission line when the load (antenna) impedance differs from the characteristic impedance of the line.

In a transmission line terminated in a pure resistance having a value equal to the characteristic impedance of the line, no standing waves occur. This is the ideal situation. When standing waves are present, a nonuniform distribution of current and voltage exists. The greater the impedance mismatch, the greater the nonuniformity. The ratio of the maximum voltage to the minimum voltage is the *voltage standing-wave ratio* (VSWR), often called simply the *standing-wave ratio* (SWR). The ratio of the maximum current to the minimum current can also be defined as the SWR, but this definition is rarely used. The ideal SWR is 1:1. Mismatches are represented as ratios such as 2:1 (maximum voltage twice the minimum voltage), 3:1 (maximum voltage 3 times the minimum), and so on.

A high SWR can sometimes be tolerated, but at high frequencies or with long transmission lines, an impedance mismatch can cause a significant loss of signal power. If a high-power transmitter is used, the currents and voltages on a severely mismatched feed line can become large enough to cause physical damage by heating up the conductors and/or the dielectric material separating them. The SWR on a transmission line can be easily measured using a commercially available meter designed for the purpose.

Safety Issues

Antenna systems can present significant physical and electrical dangers to personnel who install and maintain them. Antennas should never be placed in such a way that they can fall or blow down on power lines. Also, it should not be possible for power lines to fall or blow down on an antenna. Antennas and supporting structures should be prevented from posing a physical danger to people near them; for example, a radio tower should not be accessible to children, because they might try to climb it.

Wireless equipment having outdoor antennas should not be used during thundershowers, or when lightning is anywhere in the vicinity. Antenna construction and maintenance should never be undertaken when lightning is visible, even if a storm appears to be distant. Ideally, antennas should be disconnected from electronic equipment, and connected to a substantial earth ground, at all times when the equipment is not in use.

The climbing of antenna supports such as towers and utility poles is a job for professionals. Under no circumstances should an inexperienced person attempt to climb such a structure.

Indoor transmitting antennas expose operating personnel to EM field energy. The extent of the hazard, if any, posed by such exposure has not been firmly established. However, there is sufficient concern to warrant checking the latest publications on the topic. For detailed information concerning antenna safety, consult a professional antenna engineer and/or a comprehensive text on antenna design and construction.

Always follow this basic safety rule: If you are not confident in your ability to install or maintain an antenna system, leave it to a professional.

Quiz

Refer to the text in this chapter if necessary. A good score is eight correct. Answers are in the back of the book.

1. The efficiency of a ground-mounted vertical antenna system, including the feed line and radiating element, can be optimized by
   (a) maximizing the radiation resistance.
   (b) minimizing the loss resistance.
   (c) minimizing the SWR.
   (d) All of the above

2. In a full-size, 3-element Yagi antenna, the director is usually
   (a) shorter than the driven element.
   (b) longer than the driven element.
   (c) the same length as the driven element.
   (d) any length; it doesn't matter.

3. A quarter-wave vertical antenna, made from aluminum tubing with a velocity factor of 91%, is exactly 10 m tall. What is the natural resonant frequency of this antenna?
   (a) 68 MHz
   (b) 6.8 MHz
   (c) 136 MHz
   (d) 13.6 MHz

4. Ground loops are undesirable because
   (a) they can act as antennas and increase the risk of EMI.
   (b) they increase the risk of electric shock.

(c) they can't easily be made large enough.

(d) – Hold it! Ground loops are a good thing.

5. In a ground-plane antenna, the radiating element
    (a) should be less than a quarter wavelength long.
    (b) should be more than a quarter wavelength long.
    (c) should be resonant at the operating frequency.
    (d) should not be resonant at the operating frequency.

6. Voltage variations on an RF transmission line when the antenna impedance differs from the characteristic impedance of the line are called
    (a) radials.
    (b) short waves.
    (c) electromagnetic waves.
    (d) standing waves.

7. Assuming a constant, known loss resistance, the efficiency of an antenna
    (a) increases as the radiation resistance increases.
    (b) decreases as the radiation resistance increases.
    (c) does not change as the radiation resistance increases.
    (d) cannot be determined even if the radiation resistance is known.

8. Coaxial cable is an example of
    (a) balanced line.
    (b) unbalanced line.
    (c) parallel-wire line.
    (d) TV ribbon.

9. A $\lambda/2$ radiator, fed at the end with a $\lambda/4$ section of parallel-wire transmission line, is an example of
    (a) a doublet antenna.
    (b) a parasitic array.
    (c) a phased array.
    (d) None of the above

10. What is the end-to-end length, in feet, of a half-wave open dipole antenna cut for a frequency of 15,000 kHz? Assume that the antenna is made of wire, and the velocity factor of the wire is 95%.
    (a) 9.533 ft
    (b) 0.009533 ft
    (c) 31.13 ft
    (d) 0.03113 ft

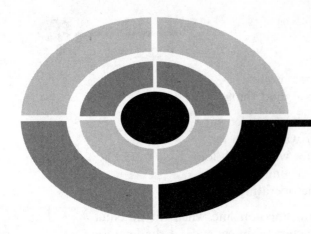

Test: Part Three

Do not refer to the text when taking this test. You may draw diagrams or use a calculator if necessary. A good score is at least 30 correct. Answers are in the back of the book. It's best to have a friend check your score the first time, so you won't memorize the answers if you want to take the test again.

1. Fill in the blanks with the word that makes the following sentence true: "In spread-spectrum communications, the transmitter _____ is rapidly varied independently of signal modulation, and the receiver _____ is programmed to automatically follow along."
   (a) detector
   (b) gain
   (c) frequency
   (d) antenna
   (e) impedance

2. Figure Test3-1 is most likely a spectral diagram of
   (a) a single-sideband signal.
   (b) a polarization-modulated signal.

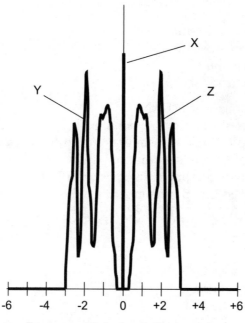

Relative amplitude

Frequency, kilohertz relative to carrier

Fig. Test3-1.   Illustration for Part Three Test Questions 2 through 5.

(c) an amplitude-modulated signal.
(d) a frequency-shift keyed signal.
(e) a continuous-wave signal.

3.   In Fig. Test3-1, X indicates
     (a) the carrier.
     (b) the lower sideband.
     (c) the upper sideband.
     (d) the deviation.
     (e) the frequency shift.

4.   In Fig. Test3-1, Y indicates
     (a) the carrier.
     (b) the lower sideband.
     (c) the upper sideband.
     (d) the deviation.
     (e) the frequency shift.

5. In Fig. Test3-1, Z indicates
    (a) the carrier.
    (b) the lower sideband.
    (c) the upper sideband.
    (d) the deviation.
    (e) the frequency shift.

6. A local area network (LAN) is
    (a) a group of radio receivers linked by wires or cables.
    (b) a group of radio transmitters linked by wireless.
    (c) a group of computers linked within a small geographic region.
    (d) a group of computers linked over a large geographic region.
    (e) any linked group of computers.

7. A half-wavelength wire fed at the center forms an antenna called
    (a) a Yagi.
    (b) a quad.
    (c) a phased array.
    (d) a parasitic array.
    (e) None of the above

8. A good example of a wide area network (WAN) is
    (a) a set of computers linked by wireless, all located in the same house.
    (b) a set of computers linked by cables, all located in the same building.
    (c) a set of computers linked by cable, all located on a single college campus.
    (d) a set of computers linked by wireless or cable, located on various corporate campuses throughout North America.
    (e) a communication between two people using amateur radio transceivers.

9. Imagine a vertical antenna mounted on the ground, with a system of radials to minimize ground loss. Suppose the vertical radiator is exactly a quarter wavelength high. If the transmission frequency is slightly lowered and all other conditions are kept constant, the radiation resistance of this antenna can be expected to
    (a) disappear.
    (b) decrease.
    (c) stay the same.
    (d) increase.
    (e) become infinite.

10. A satellite in a low-earth-orbit (LEO) network might be found
    (a) at an altitude lower than 10 kilometers.
    (b) in an orbit around the moon.
    (c) in a geostationary orbit.
    (d) at an altitude higher than 36,000 kilometers.
    (e) in an orbit that takes it over the earth's geographic poles.

11. A device that can act as both a modulator and a demodulator is called
    (a) an oscillator.
    (b) a reactor.
    (c) a modem.
    (d) a monitor.
    (e) a sideband.

12. Imagine a half-wave length of wire fed at the center. If the length of the wire is made slightly shorter while all other conditions are kept constant, the radiation resistance of this antenna can be expected to
    (a) disappear.
    (b) decrease.
    (c) stay the same.
    (d) increase.
    (e) become infinite.

13. Internet relay chat (IRC) is a scheme in which
    (a) computer users leave messages for each other on Internet bulletin boards.
    (b) computer users carry on real-time script conversations with other computer users over the Internet.
    (c) amateur radio operators communicate using geostationary satellites.
    (d) people with wireless receivers attempt to eavesdrop on Internet communications.
    (e) people send e-mail messages for advertising purposes.

14. In a superheterodyne receiver, the incoming signal from the antenna is first passed through a tunable, sensitive amplifier called the
    (a) mixer.
    (b) modulator.
    (c) detector.
    (d) local oscillator.
    (e) front end.

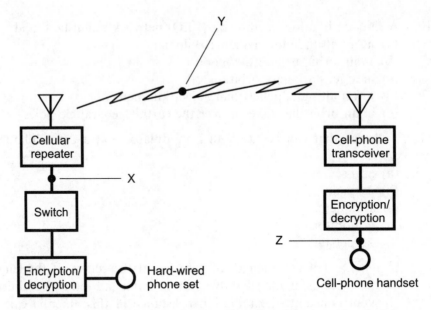

Fig. Test3-2.   Illustration for Part Three Test Questions 15 and 16.

15.   Figure Test3-2 illustrates
      (a) end-to-end encryption.
      (b) wireless-only encryption.
      (c) wiretapping.
      (d) a hard-wired telephone connection.
      (e) None of the above

16.   At what point in Fig. Test3-2 would a person with a general-coverage, wireless receiver attempt to eavesdrop on the communications?
      (a) Point X
      (b) Point Y
      (c) Point Z
      (d) Any of the points X, Y, or Z
      (e) None of the points X, Y, or Z

17.   Facsimile, also called fax, is a method of transmitting and receiving
      (a) high-fidelity audio.
      (b) high-speed Morse code.
      (c) non-moving images.
      (d) full-motion video.
      (e) control signals.

18. Synchronized communications is a specialized digital mode in which
    (a) the transmitter and receiver operate from the same antenna to minimize the interaction between the two systems.
    (b) the transmitter and receiver use the same local oscillator in order to regulate the frequencies of both systems.
    (c) the transmitter and receiver fluctuate rapidly in gain according to a programmed sequence.
    (d) the transmitter and receiver operate from a common time standard to optimize the amount of data that can be sent in a channel or band.
    (e) the transmitter and receiver operate within the same dynamic range to optimize the signal intelligibility.

19. A quad antenna is normally
    (a) omnidirectional.
    (b) bidirectional.
    (c) unidirectional.
    (d) isotropic.
    (e) a phased array.

20. Which of the following is an example of the use of a personal communications system (PCS)?
    (a) The programming of a robot
    (b) Browsing the Web
    (c) FM broadcasting
    (d) The operation of a radar set
    (e) A cell-phone conversation

21. In the output of a typical frequency-modulation voice transmitter,
    (a) the instantaneous frequency remains constant.
    (b) the instantaneous amplitude remains constant.
    (c) the carrier is switched on and off.
    (d) there can be at most 8 different digital states.
    (e) there can be at most 16 different digital states.

22. In network communications, the transmitting computer is also known as
    (a) the cathode.
    (b) the node.
    (c) the byte.
    (d) the source.
    (e) the emitter.

23. A circuit that silences a receiver when no signal is present, allowing reception of signals when they appear, is called
    (a) a gain nullifier.
    (b) a squelch.
    (c) a mixer.
    (d) a ratio detector.
    (e) a digital signal processor.

24. Figure Test3-3 shows a tuned loop antenna. The inductance of the loop, L, is 100 microhenrys (μH). The capacitor is set for a value of $C = 150$ picofarads (pF), which is the equivalent of 0.000150 microfarads (μF). The formula for the resonant frequency f_o, in megahertz (MHz), for a tuned inductance-capacitance circuit where L is the inductance in microhenrys and C is the capacitance in microfarads is given by the following formula:

$$f_o = 1 \Big/ \Big[2\pi (LC)^{1/2} \Big]$$

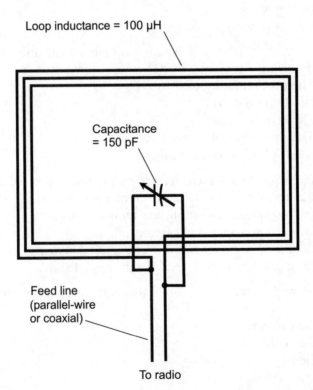

Loop inductance = 100 μH

Capacitance = 150 pF

Feed line (parallel-wire or coaxial)

To radio

Fig. Test3-3.   Illustration for Part Three Test Questions 24 through 26.

What is the resonant frequency of this tuned loop antenna? Neglect any capacitance that might exist in the feed line. Consider $\pi = 3.14$.
(a) 769 kHz
(b) 1.30 MHz
(c) 10.6 MHz
(d) There is no such frequency because the circuit is not resonant.
(e) It cannot be determined without more information.

25. Refer to Fig. Test3-3. If the variable capacitor is adjusted so its capacitance increases, the resonant frequency of the tuned antenna
(a) goes down.
(b) goes up.
(c) does not change.
(d) fluctuates alternately up and down.
(e) becomes infinite.

26. Refer to Fig. Test3-3. If more turns are added to the coil so its inductance increases, but the capacitor setting is left the same, the resonant frequency of the tuned antenna
(a) goes down.
(b) goes up.
(c) does not change.
(d) fluctuates alternately up and down.
(e) becomes infinite.

27. A quantitative measure of the amount of noise a radio receiver generates inside itself, as opposed to noise that comes from outside sources such as lightning, is known as
(a) the noise figure.
(b) the mixing efficiency.
(c) the selectivity.
(d) the gain.
(e) the dynamic range.

28. A primitive encryption scheme in which a voice signal is "flipped upside down," so the lowest audio frequencies become the highest and the highest audio frequencies become the lowest, is an example of
(a) intermodulation.
(b) audio scrambling.
(c) wireless wiretapping.
(d) digital signal processing.
(e) envelope detection.

29. Imagine a one-way communications system that employs numerous small, battery-powered wireless receivers that people keep in their pockets. Transmitters are located throughout the city. Each receiver can pick up a signal from any one of the transmitters. Such a signal causes the receiver to beep and then display a telephone number on a small screen. That's all the system can do. This is an example of
    (a) a paging system.
    (b) a wireless local loop.
    (c) wireless wiretapping.
    (d) end-to-end encryption.
    (e) e-mail.

30. Suppose a radio receiver can pick up weak signals just fine, except when there is a strong signal at a frequency near that of the desired signal. This problem suggests that the receiver has
    (a) insufficient gain.
    (b) excessive selectivity.
    (c) a poor noise figure.
    (d) poor dynamic range.
    (e) All of the above

31. In a frequency-modulated signal, the maximum extent to which the instantaneous carrier frequency differs from the unmodulated-carrier frequency is called
    (a) the deviation.
    (b) the modulation percentage.
    (c) the digital ratio.
    (d) the sampling resolution.
    (e) the sampling frequency.

32. Ideally, the reflector of a dish antenna is
    (a) perfectly flat.
    (b) shaped like two flat walls that meet in a straight line.
    (c) shaped like two flat walls and a flat ceiling that meet at a single point.
    (d) shaped like a paraboloid.
    (e) Any of the above

33. A waveguide is
    (a) a single wire along which EM fields can propagate.
    (b) a coaxial cable along which EM fields can propagate.
    (c) a hollow conduit-like structure through which EM fields can propagate.

(d) a pair of parallel wires along which EM fields can propagate.

(e) None of the above

34. The ideal standing-wave ratio (SWR) on a transmission line is
    (a) 1:0
    (b) 0:0
    (c) 1:1
    (d) 1:2
    (e) 2:1

35. In a digital signal, the number of possible states or levels is called
    (a) the sampling rate.
    (b) the sampling frequency.
    (c) the sampling frame.
    (d) the sampling resolution.
    (e) the sampling index.

36. Figure Test3-4 is a block diagram of
    (a) a crystal set.
    (b) a single-sideband transmitter.
    (c) a direct-conversion receiver.
    (d) a superheterodyne receiver.
    (e) a ratio detector.

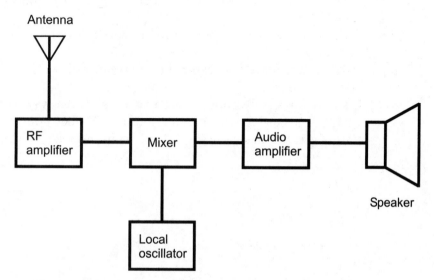

Fig. Test3-4.   Illustration for Part Three Test Questions 36 through 38.

37. Suppose, in the system shown in Fig. Test3-4, the RF amplifier gain is increased. This might be expected to produce
    (a) improved sensitivity.
    (b) reduced sensitivity.
    (c) decreased audio output.
    (d) excessive modulation.
    (e) increased deviation.

38. Suppose, in the system shown in Fig. Test3-4, the local oscillator is replaced by an amplifier. What will happen to the system as a whole?
    (a) It will become more sensitive.
    (b) It will become less sensitive.
    (c) It will become more powerful.
    (d) It will become less efficient.
    (e) It will cease to function properly.

39. Single-sideband reception requires the use of
    (a) a ratio detector.
    (b) a crystal set.
    (c) a superheterodyne receiver.
    (d) an AM receiver.
    (e) a product detector.

40. How does high-definition television (HDTV) differs from conventional TV?
    (a) HDTV portrays three dimensions, but conventional TV only two dimensions.
    (b) HDTV portrays color, but conventional TV only portrays shades of gray.
    (c) HDTV is transmitted by cable, but conventional TV is transmitted by satellite.
    (d) HDTV requires a fiberoptic cable that runs all the way to the subscriber's receiving set, but conventional TV does not.
    (e) An HDTV raster has more lines than a conventional TV raster.

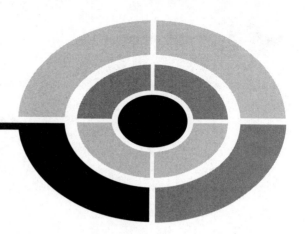

Final Exam

Do not refer to the text when taking this test. You may draw diagrams or use a calculator if necessary. A good score is at least 53 correct (75%). Answers are in the back of the book. It's best to have a friend check your score the first time, so you won't memorize the answers if you want to take the test again.

1. Susceptance in an AC circuit is expressed in
    (a) imaginary-number siemens.
    (b) imaginary-number ohms.
    (c) imaginary-number farads.
    (d) imaginary-number henrys.
    (e) imaginary-number volts.

2. Fig. Exam-1 is an illustration of
    (a) alternating current.
    (b) pulsating direct current.
    (c) a triangular wave.
    (d) a sawtooth wave.
    (e) a rectangular wave.

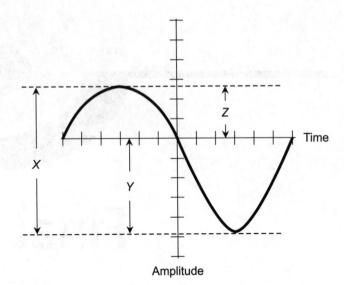

Fig. Exam-1. Illustration for Final Exam Questions 2 through 5.

3. In Fig. Exam-1, X represents the
   (a) negative peak value.
   (b) positive peak value.
   (c) peak-to-peak value.
   (d) rms value.
   (e) average value.

4. In Fig. Exam-1, Y represents the
   (a) negative peak value.
   (b) positive peak value.
   (c) peak-to-peak value.
   (d) rms value.
   (e) average value.

5. In Fig. Exam-1, Z represents the
   (a) negative peak value.
   (b) positive peak value.
   (c) peak-to-peak value.
   (d) rms value.
   (e) average value.

6. Which of the following is not an application of an NPN transistor?
   (a) Switching.
   (b) Oscillation.
   (c) Memory.

(d) Amplification.

(e) All of the above are applications of NPN transistors

7. When the output of an amplifier is fed back to the input in phase, the result is usually
(a) signal cancellation.
(b) noise limitation.
(c) oscillation.
(d) frequency modulation.
(e) amplitude modulation.

8. Imagine a voltage divider network in which there are three resistors, each with the same ohmic value, connected in series across a 12 V battery. What is the voltage across the middle resistor? Assume that the resistors have values large enough so they don't burn out or place an unreasonable strain on the battery.
(a) 2 V
(b) 3 V
(c) 4 V
(d) 6 V
(e) It is impossible to tell without more information.

9. The henry is the standard unit of
(a) phase.
(b) visible-light brightness.
(c) audio loudness.
(d) resistance.
(e) None of the above.

10. Imagine that the vertical bars in Fig. Exam-2 represent bursts of a high-frequency RF carrier wave. This graph shows a form of
(a) signal modulation.
(b) signal amplification.
(c) signal phase.
(d) signal frequency.
(e) signal efficiency.

11. The greatest average signal power in Fig. Exam-2 occurs
(a) where the vertical bars are shortest.
(b) where the vertical bars are tallest.
(c) where the vertical bars are farthest apart.
(d) where the vertical bars are closest together.
(e) nowhere; it is always zero.

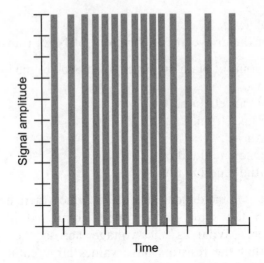

Fig. Exam-2. Illustration for Final Exam Questions 10 and 11.

12. Complex numbers are used in electronics order to define
    (a) DC resistance.
    (b) RF wavelength.
    (c) DC conductance.
    (d) RF radiation resistance.
    (e) impedance vectors.

13. A ripple filter in a power supply can consist of
    (a) an inductor in parallel with the rectifier output, and a capacitor in series with the rectifier output.
    (b) an inductor in series with the rectifier output, and a capacitor in parallel with the rectifier output.
    (c) an inductor and capacitor, both in series with the rectifier output.
    (d) a capacitor in series with the rectifier output, and a second capacitor in parallel with the rectifier output.
    (e) an inductor in series with the rectifier output, and a second inductor in parallel with the rectifier output.

14. A local oscillator is obtained to produce a constant-frequency IF signal in
    (a) a crystal radio receiver.
    (b) a Hartley circuit.
    (c) an analog-to-digital converter (ADC).
    (d) a superhet receiver.
    (e) an audio filter.

15. Which of the following specifications becomes more and more important in a radio receiver as the frequency increases?
    (a) The PA output power.
    (b) The sideband ratio.
    (c) The carrier suppression.
    (d) The noise figure.
    (e) The tracking rate.

16. Is it possible to replace a MOSFET in an electronic circuit with a JFET, and have the resulting circuit work properly?
    (a) Yes, always.
    (b) No, never.
    (c) Yes, sometimes.
    (d) Yes, but only if P-channel MOSFETs are replaced with N-channel JFETs.
    (e) Yes, but only if the source and gate electrodes are interchanged.

17. A half-wave length of wire, fed at the center and used to transmit radio signals, is called
    (a) a ground-plane antenna.
    (b) a zepp antenna.
    (c) a helical antenna.
    (d) a longwire antenna.
    (e) an open dipole antenna.

18. If a ground-plane antenna is mounted at least 1/4 wavelength above the earth's surface, how many 1/4-wave radials are sufficient to ensure reasonable efficiency?
    (a) Three or four
    (b) 36 to 48
    (c) 120 to 180
    (d) 360 or more
    (e) The antenna cannot be made efficient, no matter how many 1/4-wave radials there are.

19. An audio notch filter
    (a) cancels out AF energy having a particular frequency, allowing energy at all other audio frequencies to pass.
    (b) passes AF energy having a particular frequency, canceling out energy at all other frequencies.
    (c) passes AF energy above a certain frequency, while suppressing all energy below that frequency.

(d) passes AF energy below a certain frequency, while suppressing all energy above that frequency.

(e) passes AF energy in a range between two defined frequencies, while suppressing all energy outside that range.

20. Which of the following is a public radio communications system that allows users to operate 11-meter HF transceivers in the United States?

(a) the Global Positioning System.

(b) the Digital Satellite System.

(c) the Citizens Radio Service.

(d) the Shortwave Radio System.

(e) the American Radio Relay League.

21. An RF field-strength meter can consist of

(a) a potentiometer connected across the terminals of an ohmmeter.

(b) an FM or SSB superheterodyne transmitter.

(c) a simple DC voltmeter connected to a battery.

(d) a microammeter connected to a whip antenna through a diode.

(e) Any of the above

22. In frequency-hopping spread-spectrum communications, the dwell time is

(a) the rate at which the signals change from one mode to another.

(b) the length of time that the signal remains continuously on one frequency.

(c) equal to the bandwidth in hertz.

(d) the length of time that power is applied between signal bursts.

(e) an undefined term; there is no such thing as dwell time in frequency-hopping communications.

23. A ratio detector is a circuit designed to recover the intelligence from

(a) a CW signal.

(b) an AM signal.

(c) an FSK signal.

(d) an FM signal.

(e) an SSB signal.

24. In Fig. Exam-3, what are the potentiometers marked X meant to do?

(a) Adjust the voltage at which the signal or noise peaks are clipped.

(b) Adjust the frequency of the signal.

(c) Select the signal harmonic to be suppressed.

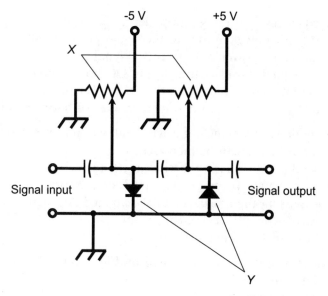

Fig. Exam-3. Illustration for Final Exam Questions 24 and 25.

(d) Adjust the extent to which current surges are suppressed.
(e) Adjust the DC output voltage.

25. In Fig. Exam-3, what are the diodes marked *Y* meant to do?
(a) Clip the peaks of the input signal or noise.
(b) Alter the frequency of the signal.
(c) Suppress signal harmonics.
(d) Suppress current surges.
(e) Increase the DC output voltage.

26. Which of the following is usually an analog mode of transmission?
(a) Radioteletype using frequency-shift keying.
(b) Morse code using CW emission.
(c) Digital television.
(d) Conventional FM.
(e) All of the above

27. The diode in a crystal radio receiver acts as
(a) a product detector.
(b) a frequency multiplier.
(c) a frequency divider.
(d) an amplitude modulator.
(e) an envelope detector.

28. The efficiency of a power amplifier is the ratio of
    (a) the DC power output to the signal power input.
    (b) the signal power output to the signal power input.
    (c) the signal power output to the DC power input.
    (d) the DC power output to the DC power input.
    (e) the collector or drain voltage to the collector or drain current.

29. The inductive reactance of a wire coil, wound on a cylindrical form
    with an air core, can be increased by
    (a) winding more turns of wire on the form.
    (b) inserting a ferromagnetic rod inside the form.
    (c) increasing the frequency of the applied AC signal.
    (d) Any of the above (a), (b), or (c)
    (e) None of the above (a), (b), or (c)

30. What sort of device is illustrated by Fig. Exam-4?
    (a) An ammeter.
    (b) A voltmeter.
    (c) A frequency meter.
    (d) An ohmmeter.
    (e) A wattmeter.

31. Suppose a 10-V battery is connected to the terminals in Fig. Exam-4
    marked "DC voltage source." Suppose the microammeter is
    deflected to full scale by this voltage (50 μA). What is the value of
    R_3, through which current flows?
    (a) 200 Ω
    (b) 2 kΩ

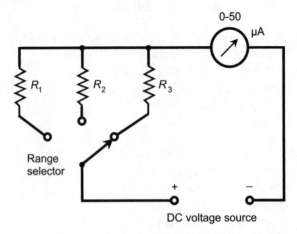

Fig. Exam-4. Illustration for Final Exam Questions 30 through 32.

(c) $200\,k\Omega$

(d) $0.2\,\Omega$

(e) $0.02\,\Omega$

32. Suppose a source of 0.05 V AC is connected to the terminals in Fig. Exam-4 marked "DC voltage source." What is the meter indication?

(a) It depends on the value of R_3.

(b) Zero, because AC is connected to a DC meter.

(c) $0.25\,\mu A$

(d) $0.1\,\mu A$

(e) The meter needle is "pinned" because it is subjected to a current higher than it is designed to register.

33. What type of antenna would most likely be used in radio direction finding (RDF) at frequencies below 300 kHz?

(a) A Yagi antenna.

(b) A quad antenna.

(c) A helical antenna.

(d) A dish antenna.

(e) A loopstick antenna.

34. The standard unit of electromotive force, also called "electrical pressure," is the

(a) ohm.

(b) coulomb.

(c) ampere.

(d) farad.

(e) None of the above

35. When a diode is forward-biased and the voltage is greater than the forward breakover voltage, the device

(a) blocks current.

(b) conducts current.

(c) generates high voltage.

(d) produces transients.

(e) acts as a capacitor.

36. What is the main difference between a circuit that uses a PNP transistor and a circuit that uses an NPN transistor?

(a) The polarities of the applied DC power-supply voltages at the electrodes are opposite in the PNP device, as compared with the NPN device.

(b) PNP devices have greater amplification potential than NPN devices.

(c) NPN devices have greater amplification potential than PNP devices.

(d) PNP devices are more stable than NPN devices.

(e) PNP devices can oscillate, but NPN devices cannot.

37. What is the difference in phase between any two AC waves in a three-phase system?
    (a) None; they are in phase.
    (b) 30°
    (c) 60°
    (d) 90°
    (e) 120°

38. Conventional AM is an inefficient mode because
    (a) it can only work at low power levels.
    (b) the audio content is too high.
    (c) the carrier consumes 2/3 of the signal power.
    (d) there is no carrier output.
    (e) it requires class C amplifiers.

39. One of the most serious problems with nickel-cadmium (NICAD) cells is the fact that
    (a) they cannot be recharged.
    (b) they have dangerously high voltages.
    (c) they explode unless connected to a load.
    (d) the cadmium they contain is toxic.
    (e) All of the above

40. A microammeter is designed to measure electrical
    (a) inductance.
    (b) voltage.
    (c) resistance.
    (d) capacitance.
    (e) None of the above

41. When the emitter-base (E-B) junction of a bipolar transistor is at zero bias and there is no signal input, the emitter-collector current is theoretically
    (a) large and negative.
    (b) large and positive.
    (c) alternating.
    (d) pulsating.
    (e) zero.

42. Five resistors, each with a value of 50 Ω, are connected in parallel. A 10 V battery is placed across this combination of resistors. What is the current drawn from the battery?
    (a) 40 mA
    (b) 200 mA
    (c) 1 A
    (d) 5 A
    (e) 25 A

43. Fill in the blank to make the this sentence true: "In an emitter follower circuit, the _____ is connected to signal ground."
    (a) gate
    (b) base
    (c) drain
    (d) source
    (e) collector

44. Which of the following is an advantage of a tuned RF power amplifier over a broadband RF power amplifier?
    (a) The tuned amplifier offers better suppression of harmonics and spurious signals.
    (b) The tuned amplifier can work at higher power levels.
    (c) The tuned amplifier is less likely to oscillate.
    (d) The tuned amplifier does not need any adjustment when the frequency is changed.
    (e) None of the above; a tuned RF power amplifier is inferior to a broadband amplifier in all respects.

45. Fill in the blank to make the following sentence true: "The _____ of an antenna is the ratio, in decibels, of the effective radiated power (ERP) in the main lobe relative to the ERP from a dipole antenna in its favored directions."
    (a) front-to-back ratio
    (b) front-to-side ratio
    (c) forward gain
    (d) power factor
    (e) noise figure

46. At which of the following frequencies would you never expect to find a Yagi antenna used at a wireless communications station?
    (a) 222.5 MHz
    (b) 50.1 MHz
    (c) 18.1 MHz

(d) 14.05 MHz

(e) 0.06 MHz

47. Which of the following types of amplifier circuit can generate significant output power, even though, in theory, it draws no power from the input signal?

(a) Class A

(b) Class AB

(c) Class B

(d) Class C

(e) Class D

48. When there is a heavy electrical storm, the best way to protect sensitive electronic equipment is to

(a) short-circuit the input to the power supply.

(b) use a fuse rated for the proper current and voltage, and make sure it is fresh.

(c) physically unplug the equipment from wall outlets until the storm has passed.

(d) make sure the circuit breakers in the utility circuit can handle the voltage.

(e) make sure your house has a two-wire electrical utility system.

49. If lightning strikes near an antenna but doesn't actually hit it, current can nevertheless be induced in the antenna as a result of

(a) electrostatic discharge.

(b) wireless wiretapping.

(c) capacitive effects.

(d) the electromagnetic pulse.

(e) inductive reactance.

50. A 1 : 1 standing-wave ratio on an antenna transmission line

(a) causes significant signal loss compared with the situation under perfectly matched conditions.

(b) causes extreme voltages and currents at various points along the line.

(c) indicates either a short circuit or an open circuit at the antenna end of the line.

(d) is the ideal situation.

(e) is impossible to achieve; theoretically such a situation cannot exist.

51. The number of magnetic flux lines passing through a square centimeter of surface area is an expression of

(a) magnetic flux density.

    (b) magnetic quantity.

    (c) magnetic polarity.

    (d) electrical voltage.

    (e) alternating current.

52. A voice SSB signal has a bandwidth of about
    (a) 2.7 Hz.
    (b) 2.7 kHz.
    (c) 2.7 MHz.
    (d) 2.7 GHz.
    (e) 270 GHz.

53. Why is an electrical ground important in an antenna system?
    (a) It can help ensure the safety of personnel using equipment connected to the system.
    (b) It can protect equipment connected to the system from damage if lightning strikes nearby.
    (c) It minimizes the risk of EMI to and from nearby electronic equipment not connected to the system.
    (d) All of the above (a), (b), and (c)
    (e) None of the above (a), (b), or (c)

54. A photovoltaic cell
    (a) produces AC directly from visible light.
    (b) produces DC directly from visible light.
    (c) converts AC into visible light.
    (d) converts DC into visible light.
    (e) converts radio waves into visible light.

55. Fill in the blank to make the following sentence true: "Diversity reception is a scheme for reducing the effects of _____ in radio reception that occurs when signals from distant transmitters are returned to the earth from the ionosphere."
    (a) digital signal processing
    (b) frequency deviation
    (c) fading
    (d) uplink interference
    (e) intermodulation distortion

56. How many diodes does a full-wave bridge circuit usually have? Assume it is not necessary to bundle diodes in series or parallel to increase the voltage rating or the current-handling capacity.
    (a) One
    (b) Two

(c) Three

(d) Four

(e) Five

57. In Fig. Exam-5, what are the devices symbolized by the circles with the lines and arrows inside?

(a) NPN bipolar transistors.

(b) PNP bipolar transistors.

(c) N-channel JFETs.

(d) P-channel JFETs.

(e) None of the above.

58. What type of circuit is shown by the diagram in Fig. Exam-5?

(a) An RF oscillator.

(b) A push-pull amplifier.

(c) A DC transformer.

(d) A DC power supply.

(e) A DC-to-AC converter.

59. What would happen if the resistor in the circuit of Fig. Exam-5 were to open up?

(a) The output signal amplitude would increase.

(b) The output signal would change from AC to DC.

(c) The output signal would change from DC to AC.

(d) The output signal would vanish, or diminish greatly.

(e) The circuit would become less stable.

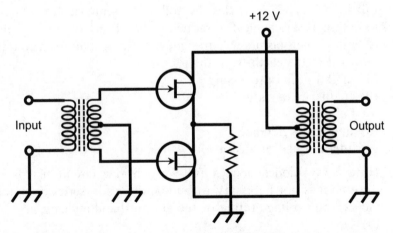

Fig. Exam-5. Illustration for Final Exam Questions 57 through 59.

60. When a semiconductor diode is used as an RF switch, it's important that
    (a) the junction capacitance be low.
    (b) the frequency be high.
    (c) the voltage be high.
    (d) the current be high.
    (e) All of the above

61. Which of the following sources is most accurate and stable for the purpose of obtaining a reference-frequency signal?
    (a) An RC Twin-T oscillator.
    (b) An LC Hartley oscillator.
    (c) A Pierce crystal oscillator.
    (d) The signal from radio station WWV.
    (e) All of the foregoing are equally accurate and stable.

62. Which of the following can be an advantage of a half-wave recti-fier circuit over a full-wave bridge rectifier circuit in certain situations?
    (a) The half-wave circuit costs less.
    (b) The half-wave circuit produces output that is easier to filter.
    (c) The half-wave circuit has better voltage regulation.
    (d) The half-wave circuit allows the use of diodes with a lower PIV rating, for the same output voltage, than does a full-wave bridge circuit.
    (e) All of the above

63. If an RF power amplifier receives an input signal that is too strong, the result can be
    (a) envelope distortion.
    (b) excessive harmonic output.
    (c) waveform distortion.
    (d) reduced efficiency.
    (e) All of the above

64. Suppose, in a DC electrical circuit, the resistance (in ohms) is symbo-lized R, the current (in amperes) is symbolized I, and the voltage (in volts) is symbolized E. Which of the following equations does not hold true according to Ohm's Law?
    (a) $I = E/R$
    (b) $R = E/I$
    (c) $E = IR$
    (d) $E = I/R$
    (e) All of the above equations hold true according to Ohm's Law.

65. The current pathway in a field-effect transistor is known as the
     (a) bias.
     (b) gate.
     (c) source.
     (d) channel.
     (e) drain.

66. In the *RX* plane, pure reactances correspond to points
     (a) on the negative-number axis.
     (b) on the real-number axis.
     (c) on the imaginary-number axis.
     (d) on neither axis.
     (e) on both axes.

67. Suppose a pair of headphones is advertised as having "an impedance
     of 1600 Ω." Which of the impedance vectors shown in Fig. Exam-6
     is meant by this expression?
     (a) Vector **A**
     (b) Vector **B**

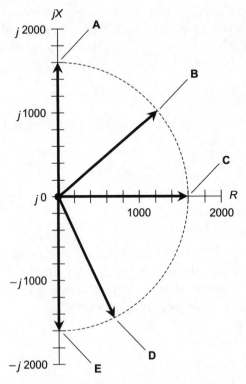

Fig. Exam-6.   Illustration for Final Exam Questions 67 and 68.

(c) Vector **C**
(d) Vector **D**
(e) Vector **E**

68. Which of the vectors in Fig. Exam-6 represents a pure capacitance?
    (a) Vector **A**
    (b) Vector **B**
    (c) Vector **C**
    (d) Vector **D**
    (e) Vector **E**

69. Theoretical DC flows from
    (a) the positive pole to the negative pole.
    (b) in the same direction as the neutrons.
    (c) from the negative pole to the positive pole.
    (d) from the south pole to the north pole.
    (e) from the north pole to the south pole.

70. The frequency of an AC wave is inversely proportional to the
    (a) amplitude.
    (b) current.
    (c) resistance.
    (d) period.
    (e) peak voltage.

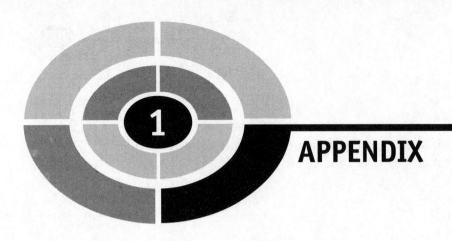

Answers to Quiz, Test, and Exam Questions

Chapter 1
1. b	2. b	3. b	4. c	5. d
6. b	7. d	8. c	9. a	10. b

Chapter 2
1. a	2. a	3. a	4. c	5. b
6. c	7. a	8. d	9. b	10. d

Chapter 3

1. d	2. a	3. a	4. b	5. a
6. b	7. c	8. d	9. c	10. d

Chapter 4

1. b	2. d	3. c	4. c	5. d
6. b	7. d	8. b	9. a	10. c

Test: Part One

1. b	2. b	3. d	4. b	5. c
6. c	7. c	8. e	9. b	10. c
11. d	12. d	13. b	14. b	15. d
16. e	17. b	18. a	19. d	20. e
21. a	22. d	23. d	24. e	25. c
26. d	27. d	28. c	29. e	30. a
31. b	32. b	33. a	34. e	35. a
36. e	37. c	38. b	39. c	40. c

Chapter 5

1. d	2. a	3. c	4. d	5. c
6. a	7. b	8. a	9. d	10. a

Chapter 6

1. d	2. d	3. c	4. b	5. c
6. d	7. a	8. c	9. d	10. c

Chapter 7

1. a	2. b	3. c	4. c	5. a
6. c	7. a	8. c	9. a	10. d

Chapter 8

1. c	2. b	3. a	4. a	5. b
6. b	7. d	8. a	9. c	10. a

Test: Part Two

1. c	2. d	3. c	4. b	5. d
6. b	7. c	8. a	9. e	10. c
11. e	12. b	13. a	14. b	15. a
16. d	17. a	18. d	19. e	20. e
21. e	22. e	23. e	24. a	25. b
26. d	27. d	28. b	29. a	30. c
31. e	32. b	33. a	34. c	35. c
36. b	37. a	38. c	39. e	40. c

Chapter 9

1. c	2. c	3. b	4. a	5. a
6. a	7. a	8. c	9. a	10. d

Chapter 10

1. c	2. d	3. b	4. b	5. d
6. a	7. d	8. a	9. c	10. d

Chapter 11

1. a	2. c	3. c	4. b	5. d
6. d	7. c	8. b	9. b	10. a

Chapter 12

1. d	2. a	3. b	4. a	5. c
6. d	7. a	8. b	9. d	10. c

Test: Part Three

1. c	2. c	3. a	4. b	5. c
6. c	7. e	8. d	9. b	10. e
11. c	12. b	13. b	14. e	15. a
16. b	17. c	18. d	19. c	20. e
21. b	22. d	23. b	24. b	25. a
26. a	27. a	28. b	29. a	30. d
31. a	32. d	33. c	34. c	35. d
36. c	37. a	38. e	39. e	40. e

Final Exam

1. a	2. a	3. c	4. a	5. b
6. c	7. c	8. c	9. e	10. a
11. d	12. e	13. b	14. d	15. d
16. c	17. e	18. a	19. a	20. c
21. d	22. b	23. d	24. a	25. a
26. d	27. e	28. c	29. d	30. b
31. c	32. b	33. e	34. e	35. b
36. a	37. e	38. c	39. d	40. e
41. e	42. c	43. e	44. a	45. c
46. e	47. a	48. c	49. d	50. d
51. a	52. b	53. d	54. b	55. c
56. d	57. c	58. b	59. d	60. a
61. d	62. a	63. e	64. d	65. d
66. c	67. c	68. e	69. a	70. d

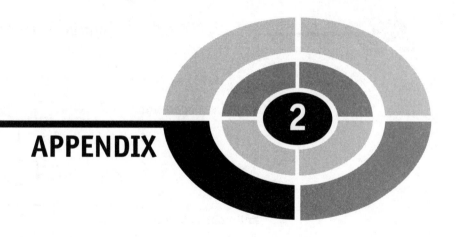

Symbols Used in Schematic Diagrams

ammeter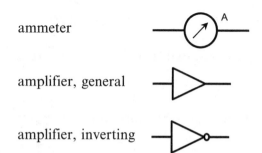

amplifier, general

amplifier, inverting

amplifier, operational

AND gate

antenna, balanced

antenna, general

antenna, loop

antenna, loop, multiturn

battery, electrochemical

capacitor, feedthrough

capacitor, fixed

capacitor, variable

capacitor, variable, split-rotor

capacitor, variable, split-stator

cathode, electron-tube, cold

cathode, electron-tube, directly heated

cathode, electron-tube indirectly heated

cavity resonator

cell, electrochemical

circuit breaker

coaxial cable

crystal, piezoelectric

delay line

diac

diode, field-effect

diode, general

diode, Gunn

diode, light-emitting

diode, photosensitive

diode, PIN

diode, Schottky

diode, tunnel

diode, varactor

diode, Zener

directional coupler

directional wattmeter

exclusive-OR gate

female contact, general

Ferrite bead

filament, electron-tube

fuse

galvanometer	
grid, electron-tube	
ground, chassis	
ground, earth	
handset	
headset, double	
headset, single	
headset, stereo	
inductor, air core	
inductor, air core, bifilar	
inductor, air core, tapped	
inductor, air core, variable	
inductor, iron core	

inductor, iron core, bifilar

inductor, iron core, tapped

inductor, iron core, variable

inductor, powdered-iron core

inductor, powdered-iron core, bifilar

inductor, powdered-iron core, tapped

inductor, powdered-iron core, variable

or

integrated circuit, general

(Part No.)

jack, coaxial or phono

jack, phone, 2-conductor

jack, phone, 3-conductor

key, telegraph

lamp, incandescent

lamp, neon

male contact, general

meter, general

microammeter

microphone

microphone, directional

milliammeter

NAND gate

negative voltage connection

NOR gate

NOT gate

optoisolator

OR gate

outlet, 2-wire, nonpolarized

outlet, 2-wire, polarized

outlet, 3-wire

outlet, 234-volt

plate, electron-tube

plug, 2-wire, nonpolarized

plug, 2-wire, polarized

plug, 3-wire

plug, 234-volt

plug, coaxial or phono

plug, phone, 2-conductor

plug, phone, 3-conductor

positive voltage connection

potentiometer

probe, radio-frequency

or

rectifier, gas-filled

rectifier, high-vacuum

rectifier, semiconductor

rectifier, silicon-controlled

relay, double-pole, double-throw

relay, double-pole, single-throw

relay, single-pole, double-throw

relay, single-pole, single-throw

resistor, fixed

resistor, preset

resistor, tapped

resonator

rheostat

saturable reactor

signal generator

solar battery

solar cell

source, constant-current

source, constant-voltage

speaker

switch, double-pole, double-throw

switch, double-pole, rotary

switch, double-pole, single-throw

switch, momentary-contact

switch, silicon-controlled

switch, single-pole, double-throw

switch, single-pole, rotary

switch, single-pole, single-throw

terminals, general, balanced

terminals, general, unbalanced

test point

thermocouple

transformer, air core

transformer, air core, step-down

transformer, air core, step-up

transformer, air core, tapped primary

transformer, air core, tapped secondary

transformer, iron core

transformer, iron core, step-down

transformer, iron core, step-up

transformer, iron core, tapped primary

transformer, iron core, tapped secondary

transformer, powdered-iron core

transformer, powdered-iron core, step-down

transformer, powdered-iron core, step-up

transformer, powdered-iron core, tapped primary

transformer, powdered-iron core, tapped secondary

transistor, bipolar, NPN

transistor, bipolar, PNP

transistor, field-effect, N-channel

transistor, field-effect, P-channel

transistor, MOS field-effect, N-channel

transistor, MOS field-effect, P-channel

transistor, photosensitive, NPN

transistor, photosensitive, PNP

transistor, photosensitive, field-effect, N-channel

transistor, photosensitive, field-effect, P-channel

transistor, unijunction

triac

tube, diode

tube, heptode

tube, hexode

tube, pentode

tube, photosensitive

tube, tetrode

tube, triode

unspecified unit or component

voltmeter

wattmeter

waveguide, circular

waveguide, flexible

waveguide, rectangular

waveguide, twisted

wires, crossing, connected

    (preferred)

or

    (alternative)

wires, crossing, not connected

    (preferred)

or

    (alternative)

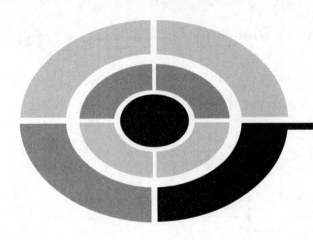

Suggested Additional References

Crowhurst, N. and Gibilisco, S., *Mastering Technical Mathematics*, 2nd edition (New York, NY: McGraw-Hill, 1999)

Dorf, R., *Electrical Engineering Handbook*, 2nd edition (Boca Raton, FL: CRC Press, 1997)

Gibilisco, S., *Handbook of Radio and Wireless Technology* (New York, NY: McGraw-Hill, 1999)

Gibilisco, S., *TAB Encyclopedia of Electronics for Technicians and Hobbyists* (New York, NY: McGraw-Hill, 1997)

Gibilisco, S., *Teach Yourself Electricity and Electronics*, 3rd edition (New York, NY: McGraw-Hill, 2002)

Horn, D., *Basic Electronics Theory with Experiments and Projects*, 4th edition (New York, NY: McGraw-Hill, 1994)

Slone, G. R., *TAB Electronics Guild to Understanding Electricity and Electronics*, 2nd edition (New York, NY: McGraw-Hill, 2000)

Van Valkenburg, M., *Reference Data for Engineers: Radio, Electronics, Computer and Communications* (Indianapolis, IN: Howard W. Sams & Co., 1998)

Veley, V., *The Benchtop Electronics Reference Manual* (New York, NY: McGraw-Hill, 1994)

INDEX

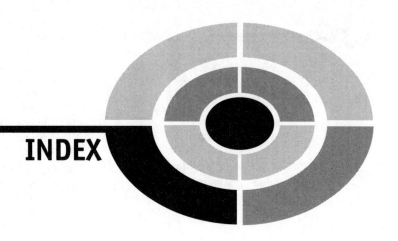

absolute amplitude, 33
absolute value, 40–1, 59
absolute-value impedance, 76
active mixer, 266
admittance, 82–8
admittances in parallel, 87–8
air-core coil, 212–13
air-variable capacitor, 212
alkaline cell, 107, 111
all-wave receiver, 294
alpha cutoff, 159
amateur radio, 146, 238, 295–6, 305
American Radio Relay League, 295
ammeter, 8–9
ampere, 3–4
ampere hour, 109
Ampere's Law, 24
ampere-turn, 23
amplification factor, 184–7
amplifier
  audio, 196–9
  bipolar, 187–8
  broadband RF power, 201
  class-A, 189–90
  class-AB, 190–1
  class-AB$_1$, 191
  class-AB$_2$, 191
  class-B, 190–2
  class-C, 190–2
  drive, 194–5
  efficiency of, 192–6
  FET, 188

  gain, 194
  IC, 188–9
  inverting, 189
  linear, 159, 190
  noise figure, 194
  operational, 178–9, 188–9
  overdrive, 194–5
  push-pull, 191–2
  RF, 199–203
  signal, 184–203
  tuned RF power, 202
  weak-signal, 200
amplitude
  absolute, 33
  expressions of, 39–40
  instantaneous, 40
  limiting, 142–3
  modulation, 192, 239–43
  negative peak, 40
  peak, 40
  peak-to-peak, 40
  positive peak, 40
  root-mean-square, 42
analog fast-scan television, 251–3, 273–4
analog-to-digital conversion, 250–1
angular frequency, 39
antenna
  billboard, 328
  broadside array, 328
  efficiency of, 311–12
  coaxial, 316–17

corner-reflector, 334–5
dipole, 312–14
dish, 332–3
doublet, 312
end-fire array, 328–9
feed line, 336–7
folded dipole, 313–14
ground system in, 320–22
ground-mounted vertical, 315–16
ground-plane, 316–17
half-wave, 312–14
helical, 332–5
horn, 146, 335
isotropic radiator, 322
large loop, 318–19
longwire, 296, 327
loopstick, 318
open dipole, 312–13
quad, 330–1
quagi, 331
quarter-wave, 314–17
random wire, 327
reference, 322
safety issues, 337–8
small loop, 317–18
Yagi, 330–2
zepp, 313–14
armature, 25
Armstrong oscillator, 207–8
ARRL, 295
ASCII, 239
aspect ratio, 251–2
asymmetrical square wave, 33
asynchronous communications,
        280
attack, 34
attenuation factor, 186–7
attenuator, 186–7, 305
audio
    amplification, 196–9
    filter, 272
    frequency, 31–2
    scrambling, 305–6
audio-frequency oscillator, 219–22
audio-frequency-shift keying, 239–40
automotive battery, 108
avalanche effect, 133–4
avalanche voltage, 134, 136

balanced line, 80
balanced modulator, 243–5
bandpass filter, 245
bandwidth, 241–2
base station, 289
battery
    automotive, 108
    electrochemical, 106
    ideal, 109–10
    lantern, 111
    lithium, 112
    nickel-cadmium, 112
    nickel-metal-hydride, 112
    transistor, 111
Baudot, 239
beat frequency, 140, 259–60
beat-frequency oscillator, 267
beeper, 291–2
beta, 158
bias, 133, 136, 155–6, 165–7
bidirectional pattern, 326
billboard antenna, 328
binary digit, 238
binary digital mode, 238
biometric security system, 304
bipolar amplifier, 186–7
bipolar transistor, 153–63
birdies, 261
bit, 238, 283
blackout, 114
blocking capacitor, 217–18
bridge, 95–7
broadband RF power amplifier, 201
broadside array, 328
brownout, 114
buffer, 162
byte, 283

capacitive reactance, 68–71
capacitor
    air-variable, 212
    blocking, 217–18
    definition of, 68
    electrolytic, 99
    filter, 99–101
    polystyrene, 212
    silver-mica, 212
capacitor-input filter, 100–1

carrier, 238, 267
carrier frequency, 259–60
Cassegrain dish feed, 332–3
cat whisker, 137
catastrophic interference, 279
cell
    alkaline, 107, 111
    dry, 107
    electrochemical, 106
    ideal, 109–10
    in telecommunications system, 289–91
    lead-acid, 106–7, 112
    lithium, 107, 111
    mercuric oxide, 111
    mercury, 111
    nickel-cadmium, 112
    nickel-metal-hydride, 112
    phone, 289–91, 304
    primary, 107
    secondary, 107
    silver-oxide, 111
    spacecraft, 112
    standard, 108
    Weston standard, 108
    zinc-carbon, 107, 111
cell-phone cloning, 304
cellular telecommunications, 289–91
channel, in FET, 163
character, 238
characteristic curve, 156–7, 168, 171
characteristic impedance, 79–82
charge carrier, 3–4, 23
charge-storage diode, 139
choke-input filter, 100–1
cipher, 300, 306
circuit breaker, 105–6
Citizens Radio Service, 294
Clapp oscillator, 211–12
class-A amplifier, 189–90
class-AB amplifier, 190–1
class-AB$_1$ amplifier, 191
class-AB$_2$ amplifier, 191
class-B amplifier, 190–2
class-C amplifier, 190–2
clickable link, 285
client-server LAN, 286
clipping, 136, 143
clipping voltage, 136

cloning, 304
closed-loop configuration, 178–9
cloud-to-ground lightning, 296–7
coaxial antenna, 316–17
coaxial cable, 80
coherent communications, 276–7
coherent radiation, 148
coil, air-core, 212–13
collinear corner-reflector array, 334
Colpitts oscillator, 209–11
common base, 161–2
common collector, 162–3
common drain, 174–5
common emitter, 160–1
common gate, 173–4
common source, 172–3
communications satellite, 287–9
comparator IC, 180
complex number, 56–60
complex number plane, 57
complex vector, 58–9
complex wave, 34, 36–8
component frequency, 33
condenser, 68
conductance, 12, 83–5
conductor, 4
conventional dish feed, 332–3
cordless phone, 303–4
core permeability, 213
corner-reflector antenna, 334–5
coulomb, 4
counterpoise, 322
coupling
    transformer, 198–9
    tuned-circuit, 199
crystal (diode), 259
crystal-controlled oscillator, 214–19
crystal oven, 214
crystal radio receiver, 137, 258–9, 261
crystal set, 258–9, 261
current
    direct, 3–28
    equalizing resistor, 104
    gain, 158–60, 185–6
    theoretical, 24

dash, in Morse code, 238, 306
data encryption, 279

DC-to-DC converter, 113–14
decay, 34
decibel, 82, 184–7, 322
declining discharge curve, 110
decoder, 278
decryption key, 301, 306
degrees of phase, 39
degrees per second, 39
demodulator, 259
demultiplexer, 180
depletion, 165
depletion mode, 171–2
depletion-mode MOSFET, 172
depletion region, 133, 144, 165
destination, 283
detection
  discriminator, 269
  envelope, 137, 267
  of AM, 267
  of CW, 267–8
  of FM, 268–70
  of FSK, 268
  of SSB, 269–71
  phase-locked loop, 268–9
  product, 269–71
  ratio, 269–70
  slope, 268–9
deviation, 247–8
diamagnetic material, 25
dielectric, 80, 133
digital
  encryption, 301
  integrated circuit, 176
  Satellite System, 274
  signal processing, 263, 277–8
  television, 274–6
digital-to-analog converter, 274–5
diode
  charge-storage, 139
  Esaki, 147
  Gunn, 146
  hot-carrier, 139–40
  IMPATT, 147
  infrared emitting, 147–9
  laser, 148
  light emitting, 147–9
  oscillator, 211
  photodiode, 149

photosensitive, 147, 149
  PIN, 138
  P-N junction in, 131–6
  point-contact, 137
  rectifier, 134
  signal, 136–46
  silicon, 133
  snap, 139
  step-recovery, 139
  tunnel, 147
  varactor, 133, 145
  Zener, 135
dip, power, 114
dipole antenna, 312–14
direct-conversion receiver, 259–60
direct current
  circuits, 2–38
  component, 40–2
  error voltage, 219
  generator, 6
  magnetism, 23–8
  power input, 193
  pulsating, 6–7, 40
  pure, 6–7
  sources of, 6
  transformer, 113–14
direction of vector, 46, 58–9
directivity plot, 322–4
director, 329
discharge curve, 109–10
discrete components, 176
discriminator, 269
dish antenna, 332–3
disk, magnetic, 27–8
diversity reception, 276–7
domain name, 285
domain type, 285
dot, in Morse code, 238, 306
double balanced mixer, 141–2
double-conversion superhet, 261
doublet antenna, 312
downlink, 274–6
drain, in FET, 163
drive, 194–5
driven element, 325
drum, 251
dry cell, 107
dual-diversity reception, 276–7

dual potentiometer, 144
dummy antenna, 305
dwell frequency, 279
dwell time, 279
dynamic current amplification, 158–9
dynamic mutual conductance, 168–9
dynamic range, 262, 264

efficiency, of amplifier, 192–6
electric field, 71
electric flux, 25
electric generator, 49
electrical grounding, 320–21
electrically erasable programmable
       ROM, 181
electrochemical
   cell, 6, 106
   energy, 106–7
   power sources, 106–12
electrolytic capacitor, 99
electromagnet, 23
electromagnetic
   coupling, 320
   field, 51
   interference, 320
   noise, 33
   pulse, 136
   radiation loss, 51
electromotive force, 6
electron, 3–4
electronic mail, 285, 291–2
electronic switching, 137–8
electron-tube voltage regulator, 102
electrostatic discharge, 171
electrostatic shielding, 25
elevation plane, 323–4
EM radiation loss, 51
e-mail, 285, 291–2
emission, 237
emitter follower, 162–3
encoding, 278, 306
encryption, 279, 301–2, 306
end-fire array, 327–8
end-to-end encryption, 301–2
energy, 49
enhancement-mode MOSFET, 172
envelope detection, 137, 267
epitaxial layer, 140

epitaxy, 140
erasable programmable ROM, 181
Esaki diode, 147

facsimile, 251, 292–3
fading, 276
farad, 68
fast-scan television, 251–3, 273–6
fax, 251, 292–3
feed line
   balanced, 336
   unbalanced, 336
   standing waves on, 337
feedback, 161–2, 189, 206–7
feedback loop, 206–7, 219
ferromagnetic material, 25
FET amplifier, 188
fiberoptic technology, 149
field-effect transistor, 163–76, 200
file server, 286
file transfer protocol, 285
filter
   bandpass, 245
   capacitor, 99–101
   choke, 99–101
   in power supply, 99–101
firmware, 181
first intermediate frequency, 260, 266
flat discharge curve, 110
flat topping, 194
flux density, 23
flux line, 23
focal point, 332
folded dipole antenna, 313–14
forward bias, 132–3, 136, 156
forward breakover, 133–4, 136, 167
forward gain, 324
fossil-fuel energy, 49
frame, 251
free space, 25
frequency
   angular, 39
   control, 144–5
   definition of, 31–3
   divider, 277
   domain, 36
   doubler, 249
   hopping, 279

modulation, 192, 245–8
multiplication, 138–9, 247
multiplier, 247
response, 196–7
shift keying, 238–40
sweeping, 280
synthesizer, 218–19
trimming, 216
front end, 200, 260, 262, 264–5
front-to-back ratio, 324–5
front-to-side ratio, 324–5
full-wave bridge, 95–7
full-wave center-tap rectifier, 95–6
fundamental frequency, 33, 138
fuse, 105

gain, 185–7, 194, 200
gain bandwidth product, 159
gain-versus-frequency curve, 196–8
galena, 137
gallium arsenide, 145, 200
gate, in FET, 163
gauss, 23
GB plane, 85–6
General Mobile Radio Service, 294
general-coverage receiver, 294
generating plant, 49–50
geostationary satellite, 287–9
geothermal energy, 49
gigahertz, 32
gilbert, 23
Global Positioning System, 288
GMRS, 294
graphic equalizer, 196
grayscale, 253
ground
  bus, 321
  loop, 321
  radial system, 315
  rod, 103
  system, 320–22
grounding
  definition of, 103
  electrical, 320
  RF, 321–2
ground-plane antenna, 316–17
ground-to-cloud lightning, 296–7
Gunn diode, 146

Gunn effect, 146
Gunnplexer, 146

half-wave antenna, 312–14
half-wave rectifier, 94–6
"ham" radio, 146, 238, 295–6
harmonic, 33, 36–7, 138, 191
hand-print recognition, 304
handy-talkie, 295
Hartley oscillator, 207–9
header, 282–3
helical antenna, 332–5
henry, 61
hertz, 32
heterodyne, 272
high-definition television, 251, 254–5
high frequency, 293
high-level AM, 240–2
highpass response, 179
high-tension line, 50–1
hobby communications, 293–6
horizontal plane, 323–4
horizontal synchronization, 253
horn antenna, 146, 335
hot-carrier diode, 139–40
hydroelectric energy, 49
hydroelectric power plant, 50
hypotenuse, 59

icon, 285
ideal battery, 109–10
ideal cell, 109–10
IF stages, 262–3, 266
IGFET, 170
image
  signals in superheterodyne
    receiver, 261
  rejection, 266
  resolution, 254
  transmission, 251–5
imaginary number, 56–60
imaginary number line, 57
impact avalanche transit time, 147
IMPATT diode, 147
impedance, 56–88
impedance mismatch, 81
impedances in series, 77–8
inductive loading, 315

inductive reactance, 60–3
inductor, 60–1
infrared emitting diode, 147–9
input port, 141
input-to-output voltage ratio, 92
instantaneous amplitude, 40
insulated-gate FET, 170
insulator, 4
integrated circuit, 102, 176–81
interactive PV system, 50, 115
intercloud lightning, 296–7
interlacing, 254
intermodulation distortion, 264
Internet, 282–7
Internet relay chat, 285
Internet telephone, 285
interrupt signal, 115
interruption, power, 114
intracloud lightning, 296–7
intrinsic semiconductor, 138
inverter, power, 113–15
inverting amplifier, 189
inverting input, 178
iris-print recognition, 304
irregular wave, 37
isotropic radiator, 322

j operator, 56
JFET, 163–9
junction capacitance, 133
junction FET, 163–9

kilohertz, 32
kilohm, 4
kilovolt, 6
kilowatt hour, 108
Kirchhoff's Laws, 18–19

L section, 100–1
lagging phase, 45–7
lantern battery, 111
laser diode, 148
lasing, 148
latency, 288
lead-acid cell, 106–7, 112
leading edge, 34
leading phase, 45–6
level-0 security, 300

level-1 security, 300–1
level-2 security, 301
level-3 security, 301
light emitting diode, 147–9
lightning, 296–9
lightning arrestor, 298
lightning rod, 298
limiter, 268–9
linear amplifier, 159
linear IC, 176
lithium battery, 112
lithium cell, 107, 111
load, 51
loading coil, 315
loading control, 202
local area network, 286–7
local oscillator, 140, 259
logarithm, 185
longwave band, 294
longwire antenna, 296, 327
loop, 80
loop antenna
    large, 318–19
    small, 317–18
loopstick antenna, 318
loss, 185
loss resistance, 311
low-earth-orbit satellite, 287–8
low-frequency band, 294
low-level AM, 240–2
lowpass response, 179

magnetic disk, 27–8
magnetic field, 23, 63
magnetic tape, 27
magnetism, 23–28
magnetomotive force, 23
magnitude of vector, 46, 58–9
main lobe, 324–5
major lobe, 327
majority carrier, 164
mark, 238
maximum deliverable current, 109
maxwell, 23
medium-frequency band, 294
medium-wave band, 294
megahertz, 32
megohm, 4

memory, 180–1
memory backup, 180
mercuric oxide cell, 111
mercury cell, 111
metal-oxide-semiconductor field-effect
        transistor, 169–72
microampere, 4
microfarad, 69
microhenry, 61
microvolt, 6
milliampere, 4
millifarad, 69
millihenry, 61
millivolt, 6
mil-spec security, 301
minor lobe, 327
mixer, 140–2, 145–6, 259, 262, 265–6
mixing product, 140, 265
modem, 239, 284–5
modular construction, 177–8
modulation
    amplitude, 239–43
    analog fast-scan television, 251–3
    audio-frequency-shift keying, 239
    emission, 237
    envelope, 191
    facsimile, 251
    fast-scan television, 251–3
    fax, 251
    frequency, 245–8
    frequency-shift keying, 238–40
    index, 247–8
    on/off keying, 238–9
    percentage, 241
    phase, 246–7, 249
    pulse, 248–9
    radioteletype, 239–40
    reactance, 245–6
    single sideband, 243–5
    television, 251–5
    ultra high-definition video, 254
modulator, balanced, 243–5
modulator/demodulator, 239
Morse code, 238, 306
MOSFET, 169–72
multiplexing, 180, 278–9
multiplexer/demultiplexer, 180
multi-section filter, 100–1

multivibrator, 221–2
musical instrument digital interface,
        220

nanoampere, 4
narrowband FM, 247
National Institute of Standards and
        Technology, 219
n-by-n matrix, 14
N-channel JFET, 164–5
N-channel MOSFET, 170
negative degree, 71
negative feedback, 189
negative peak amplitude, 40
negative resistance, 146
Net traffic, 285
network, 282–7
nickel-cadmium battery, 112
nickel-cadmium cell, 112
nickel-metal-hydride battery, 112
nickel-metal hydride cell, 112
node
    electrical, 81
    in communications, 283
noise figure, 194, 200, 262
noise limiter, 143–4
noise limiting, 143–4
noninverting input, 178
nonlinear circuit, 159
nonlinearity, 138, 159
nonreactive impedance, 76
nonvoltatile memory, 180
normally closed relay, 26
normally open relay, 26
NPN transistor, 154–7
nuclear energy, 49

ohm, 4
ohmic resistance, 50
Ohm's Law, 7–10
on/off keying, 238–9
open dipole antenna, 312–13
open-loop configuration, 178–9
operational amplifier, 178–9, 188–9
optical coupler, 149
optical scanner, 251, 293
optoisolator, 149
oscillation, 161, 201, 206–22

oscillator
  Armstrong, 207–8
  audio-frequency, 219–22
  Clapp, 211–12
  Colpitts, 209–11
  crystal-controlled, 214–19
  diode, 211
  Hartley, 207–9
  multivibrator, 221–2
  Pierce, 214–15
  reference, 219
  Reinartz crystal, 215–16
  reliability, 213
  RF, 206–212
  stability, 211–14
  twin-T, 220–1
  variable-frequency crystal, 216–17
  voltage-controlled, 217–18
output port, 141
overdrive, 194–5

packet, 282–4
packet radio, 295
pager, 291–2
paraboloidal reflector, 332
parallel-wire line, 80
parasitic array, 329–332
parasitic element, 329
passband, 262
passive mixer, 266
P-channel JFET, 164–5
P-channel MOSFET, 170
peak amplitude, 40
peak inverse voltage, 94, 134
peak reverse voltage, 94
peak-to-peak amplitude, 40
peer-to-peer LAN, 286
perfectly conducting ground, 310
period, 31–2
periodic AC wave, 31–2
permeability, 25–6, 213
personal communications systems, 289–93
phase
  angle, 64–7, 72–5
  coincidence, 43–4
  comparator, 218–19
  degrees of, 39
  lagging, 45–7

leading, 45–6
locked loop, 218–19, 268–9
modulation, 246–7, 249
of AC, 52
opposition, 44–5
radians of, 39
relationships, 43–9
phased array, 325–9
phasing harness, 326–7, 329
photodetector, 251
photodiode, 149
photoemission, 147
photosensitive diode, 147, 149
photovoltaic cell, 6, 50, 115, 149–50
photovoltaic effect, 149
photovoltaic energy-generating
    system, 50
Pierce oscillator, 214–15
piezoelectric crystal, 214
personal identification number, 304
PIN diode, 138
pinchoff, 165
pi section, 101
P-N junction, 131–6
PNP transistor, 154, 157–8
point-contact diode, 137
polar orbit, 288
polarization, 103
polystyrene capacitor, 212
positive peak amplitude, 40
post-detector stages, 263
potential difference, 16
potentiometer, 8–9, 144
power
  definition of, 10
  distribution of, 17–18
  division of, 13
  effective radiated, 322
  gain, 185–7, 322
  grid, 115
  input, 193
  inverter, 50, 113–15
  output, 193
  supplies, 91–115
  transistor, 102
preamplifier, 264
prefix multiplier, 4–5
preselector, 262, 265

primary cell, 107
primary-to-secondary turns
     ratio, 92
programmable divider, 218
pulsating DC, 94
pulse modulation, 248–9
pure resistance, 76
push-pull amplifier, 191–2
Pythagorean theorem, 59

quad antenna, 330–1
quagi antenna, 331
quarter-wave antenna, 314–17
quartz crystal, 214
quick-break fuse, 105

radians of phase, 39
radians per second, 39
radiation resistance, 309–12
Radio Emergency Associated
     Communications Teams, 294
radio direction finding, 317–18
radio frequency
     definition of, 32
     receiver, 258–80
     transmitter, 237–55
radiotelegraphy, 192
radioteletype, 192, 239–40
ramp, 34–5
random-access memory, 180
random wire, 327
raster, 254
ratio detector, 269–70
RC circuit, 71–9
REACT, 294
reactance
     capacitive, 68–71
     inductive, 60–3
     modulation, 245–6
read-only memory, 180–1
real number line, 57
receiver, 258–80
     alignment, 261
     crystal set, 258–9, 261
     direct-conversion, 259–60
     specifications, 261–2
     superheterodyne, 260–1, 262–3
     television, 273–6

rectifier
     full-wave bridge, 95–7
     full-wave center-tap, 95–6
     half-wave, 94–6
reference antenna, 322
reference oscillator, 219
reflector, 329
Reinartz crystal oscillator, 215–16
relay, 25–7
     normally closed, 26
     normally open, 26
rectangular response, 272
rectangular wave, 33
rectification, 134
rectifier diode, 134
release, 34
repeater, 289
resistance
     definition of 4–5
     in network, 11–22
     in parallel, 11–13, 15–17
     in series, 11–12, 15
     in series-parallel, 13–15
resistive impedance, 76
resistive network, 11–22
resolution
     image, 254
     sampling, 250
resonance, 78, 88, 199
resonant circuit, 78
resonant frequency, 78–9
resonant notch, 179
resonant peak, 179
reverse bias, 132–3, 136, 156
reverse parallel, 142
RF
     amplification, 199–203
     choke, 211
     grounding, 321–2
     spectrum, 199
right-hand rule, 24
ripple, 99
rise, 34
RL circuit, 63–7
root-mean-square amplitude, 42
RX plane, 75–7
RX$_C$ plane, 69–71
RX$_L$ plane, 62–3

sampling, 250–1
satellite, 287–9
saturation, 147, 156–7
sawtooth wave, 34–6
scalar, 58
scan converter, 275
scanner, 251, 293
schematic diagram, 7
second intermediate frequency,
        260–1, 266
secondary cell, 107
security, levels of, 299–301
selective squelching, 273
selectivity, 260, 262, 266
semiconductor diode, 94
sensitivity, 200, 261–2
series-parallel resistive network, 13–15
shape factor, 266
shelf life, 109
shortwave band, 276, 293–6
shortwave listening, 293–6
shortwave receiver, 293
shield, 80
side lobe, 324–5
sideband
   lower, 241–5
   upper, 241–5
siemens, 12, 83
signal diode, 136–46
signal power output, 193
signal-plus-noise-to-noise ratio, 261–2
signal-to-noise ratio, 261–2
silicon diode, 133
silver-mica capacitor, 212
silver-oxide cell, 111
sine wave, 32–3, 238
single-conversion superhet, 260–3
single balanced mixer, 141
single sideband, 243–5
skirt, 268
slope detection, 268–70
slow-blow fuse, 105
slow-scan television, 251, 253, 275
slug, 211
snap diode, 139
solar battery, 150
solar cell, 115, 150
solar energy, 49

solar panel, 150
solar-electric power supply, 115
solenoidal coil, 25
solenoidal core, 222
source
   follower, 174–5
   in communications, 282–3
   in FET, 163, 174–5
space, 238
spacecraft cell, 112
spectrum analyzer, 35–38
spherical reflector, 332
spread spectrum, 279, 304–5
spurious emissions, 201
square wave
   asymmetrical, 33
   symmetrical, 33
   theoretically perfect, 34
squelch, 272–3
stability, of oscillator, 211–14
stand-alone PV system, 50, 115
standard cell, 108
standing wave, 80–2, 337
standing-wave ratio, 82, 337
static electricity, 6
static forward current transfer
        ratio, 158–9
step-down transformer, 51–2, 91–3
step-recovery diode, 139
step-up transformer, 91–3
storage capacity, 108–9
substrate, 148, 164
surge current, 98, 103–4
surge impedance, 79
superheterodyne receiver, 260–1
susceptance, 84–5
switching, electronic, 137–8
symmetrical square wave, 33
synchronized communications, 276–7

T section, 101
telecommunications, 282–306
television
   reception, 273–6
   transmission, 251–5
tape, magnetic, 27
terahertz, 32
terminal unit, 239

tesla, 23
theoretical current, 24
three-phase AC, 52
three-wire AC system, 103
thyrector, 135–6
time-domain display, 33
timer IC, 180
tone control, 196
tone squelching, 304
toroidal core, 222
tracking, 265
trailing edge, 34
transconductance, 168–9
transformer
  coupling, 198–9
  DC, 113–14
  ratings, 93
  step-down, 51–2, 91–3
  step-up, 91–3
transient, 104–5, 135–6
transient suppressor, 104–5, 135–6, 298
transistor
  battery, 111
  bipolar, 153–63
  field-effect, 163–76
  NPN, 154–7
  PNP, 154, 157–8
transmission line, 79–82
transmitter, 237–55
trap, 315
triangular wave, 34, 36
tuned RF power amplifier, 202
tuned-circuit coupling, 199
tuning control, 202
tunnel diode, 147
turbine, 49
twin-T oscillator, 220–1
two-wire AC system, 103

ultra high-definition video, 254
unbalanced line, 80
unidirectional pattern, 326
uninterruptible power supply, 114–15
unit imaginary number, 56
uplink, 274–6
user identification, 304
username, 285
utility power transmission, 49–53

varactor, 133, 145, 217–19, 245–6
variable-frequency crystal oscillator, 216–17
varicap, 145
vector, 46, 58–9, 62–4, 70–1, 86
vector diagrams, 46–8
velocity factor, 312
vertical synchronization, 253
voice mail, 291
voice-pattern recognition, 304
volatile memory, 180
voltage
  amplification, 163
  controlled oscillator, 145, 217–18
  definition of, 6
  divider, 18–22
  doubler, 97–8
  gain, 185–6
  multiplier, 97–8
  regulation, 97, 102, 179
  regulator IC, 179
  spike, 104
volt-ampere capacity, 93
voltmeter, 8–9
volume control, 198

walkie-talkie, 295
watt, 10
watt hour, 108
wattage, 17
waveform, 220
waveguide, 332, 336
weak-signal amplifier, 200
weber, 23
Weston standard cell, 108
wide area network, 287
wideband FM, 247
wind-driven electric power plant, 50
wind energy, 49
wire-equivalent security, 300–1
wireless
  eavesdropping, 299–300
  fax, 292–3
  local loop, 291
  tap, 299, 303
wireless-only encryption, 301–2
wiretapping, 299
World Wide Web, 285

word, 238
WWV, 219, 277, 279
WWVH, 219

Yagi antenna, 330–2

Zener diode, 102, 135
zepp antenna, 313–14
zero beat, 260
zero bias, 136, 155
zinc-carbon cell, 107, 111

ABOUT THE AUTHOR

Stan Gibilisco is one of McGraw-Hill's most prolific and popular authors. His clear, reader-friendly writing style makes his electronics books accessible to a wide audience, and his background in mathematics and research makes him an ideal editor for tutorials and professional handbooks. He is the author of *The TAB Encyclopedia of Electronics for Technicians and Hobbyists*; *Teach Yourself Electricity and Electronics*; and *The Illustrated Dictionary of Electronics*. *Booklist* named his *McGraw-Hill Encyclopedia of Personal Computing* a "Best Reference of 1996."